Big Guns of the Twentieth Century and their part in great battles

Big guns have dominated all the major battle-grounds of the twentieth century, both on land and at sea. In this graphic study, Kenneth Allen has reconstructed nine key conflicts to explain how these mighty machines have decided the course of recent history. The big guns dominated Tsushima, the western front in World War One, Jutland, Coronel and Falkland and the River Plate, and all the main battles of the Second World War from Europe to the Pacific. And even today, the giant missile-launchers patrolling the world's frontiers may once again hold its fate in their balance.

With some 150 illustrations, maps and diagrams, *Big Guns* makes the ideal introductory companion to this important field of military history.

A giant British railway gun on active service in 1941.

A battery of rail-mounted French howitzers of the Second World War (front and back endpapers).

Other books by the same author

The Battle of the Atlantic
The Wars of the Roses
The Story of Gunpowder
Soldiers in Battle
Sailors in Battle

First published 1976

49 Lansdowne Place, Hove, East Sussex BN3 1HS.

Text set in 12/14 pt Photon Baskerville, printed by photolithography, and bound in Great Britain at The Pitman Press, Bath

The famous monster German railway gun, "Dora", which did so much devastation against Russia at Sebastopol.

BIG GUNS

Of the Twentieth Century and Their Part in Great Battles

KENNETH ALLEN

FIREFLY BOOKS

A Russian anti-aircraft battery in action against German planes at the siege of Leningrad.

Right Berthold Schwarz, the German Franciscan monk said to have invented gunpowder.

Below Very early German mortars (about 1575), fitted with a simple range mechanism. They lobbed their heavy shot in a short high trajectory, and were much used in siege warfare.

Chapter One

Tsushima 1905

Practical Demonstration

By the beginning of the twentieth century, guns were nearly six centuries old. The first record of their use is found in a manuscript belonging to the city of Ghent; it states that in 1313 a German monk called Berthold Schwartz, living in Flanders, fired the first shot from a gun using gunpowder as a propellant.

Many people claimed to have invented gunpowder. But today it is generally accepted that the first man to discover the tremendous power locked up in a simple formula—charcoal, sulphur and saltpetre—was the English friar and alchemist, Roger Bacon. Bacon concealed what he had found in strange cryptic writings, which suggest that he made his discovery in the middle of the thirteenth century.

It was Berthold Schwarz, however, who saw the possibilities of this powder as a new weapon; soon the city of Ghent was making and exporting guns all over Europe. The strange shape of these first guns gave them the name of *pot de fer* (literally "iron pot"). The projectile was rammed down the muzzle of the gun; gunpowder was placed at the other end and ignited by means of a hot iron applied to a touch-hole. This method, more or less, stayed in use for more than five hundred years.

Crècy, in 1346, was the first recorded battle in which cannon were used "which with fire throw little balls to frighten and destroy horses". Another century passed before they did much real damage to the enemy. All too often they exploded, killing their own gunners. Even a king, James II of Scotland, was killed when he stood too near to a gun at the siege of Roxburgh in 1460.

A great step forward came in England in 1509 when guns were first cast in one piece from wrought iron. Before, the cannon had been made from bundles of iron, or even wooden staves, placed side by side round a core and bound in place by iron hoops; it was a method rather like the one used to make wine barrels. Hence the tube-like part of the gun was called the "barrel". A later method was to fit some iron bars together and hold them in position with broad hoops shrunk around them.

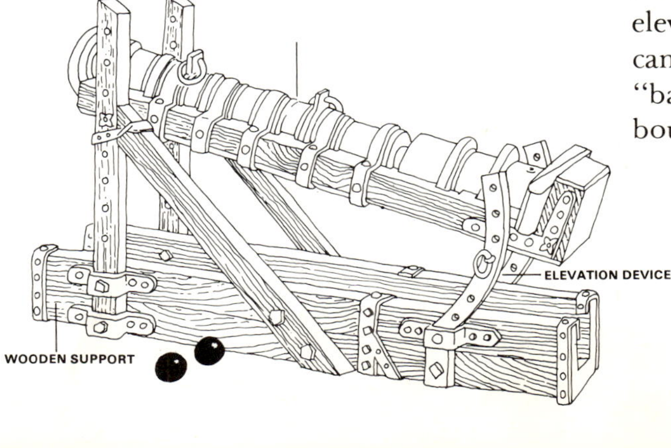

An early cannon with elevation device. The cannon was literally a "barrel" of iron staves bound with iron hoops.

The weakness of the guns was that they were muzzle-loaders; the gunsmiths of the day were unable to make a satisfactory breech that was proof against the leakage of gas from the explosion. But sometimes a hollow metal cylinder, closed at one end, was driven into the breech-end of the barrel to serve as a powder chamber. Thus the enclosed end of this cylinder became the breech of a muzzle-loading cannon.

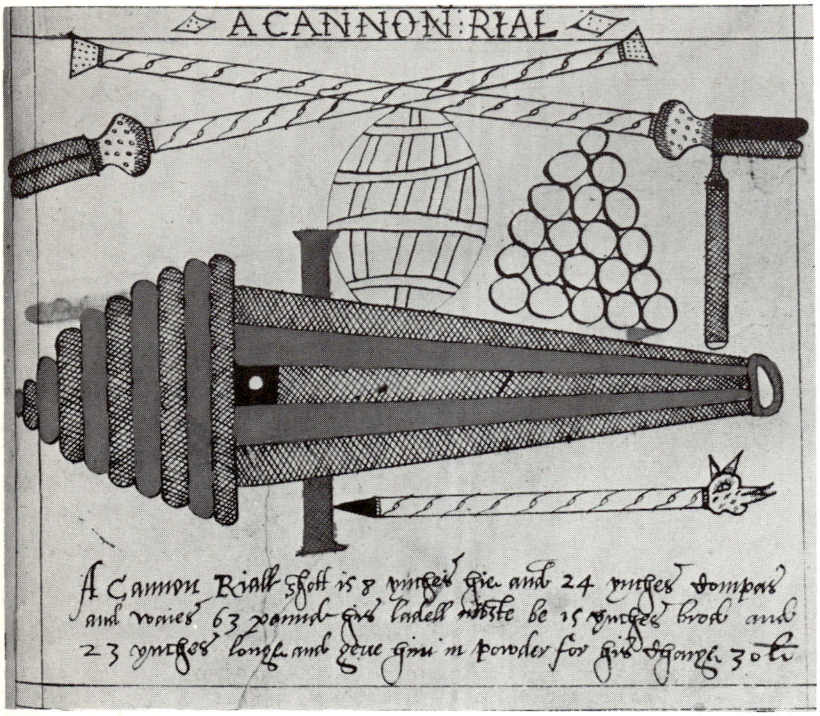

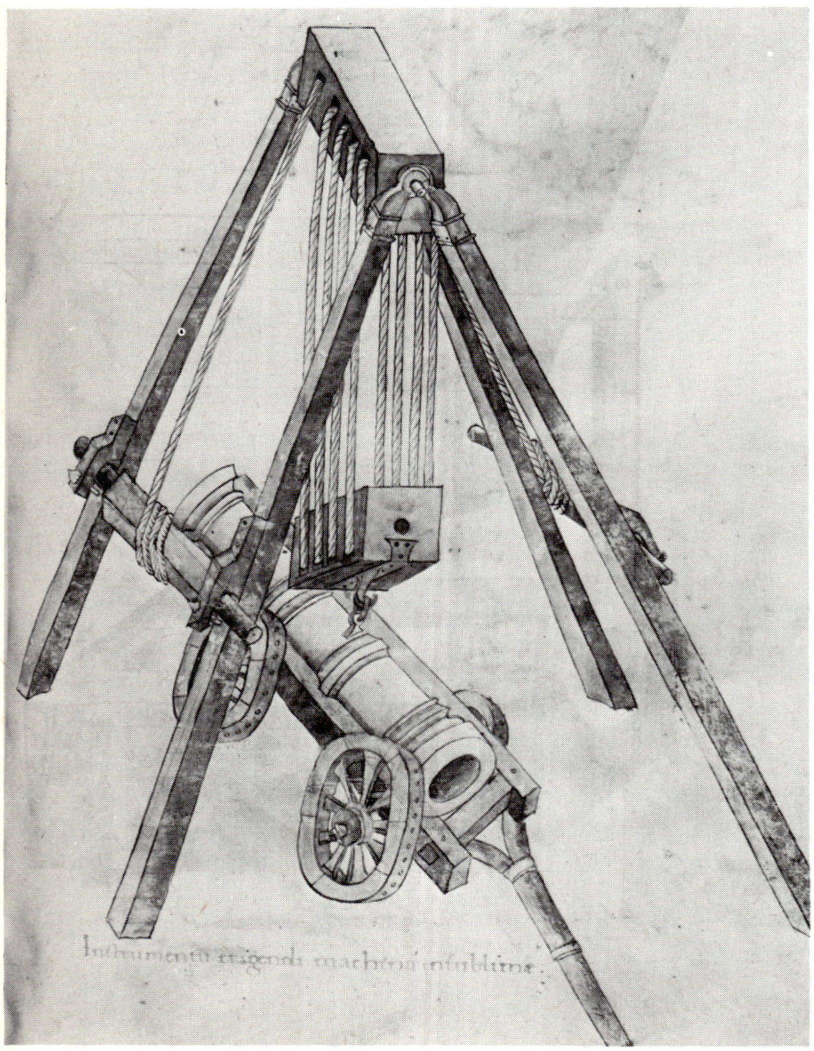

Left top A 17th-century English "cannon rial", with shot, powder and ramrods.
Left below A cannon hoist of the 15th century.
Below Typical trajectory of an early siege gun. Smoke pours from touch hole and muzzle.
Bottom Early cannon as used at Crecy.

It was Henry VIII (1509–45), a very artillery-minded king, who raised the standard of guns and gunnery in England. Henry bought a number of guns from the famous Flemish gun-maker, Hans Poppenruyter, including a set of twelve large guns known as the "Twelve Apostles". He sent some of these to oppose the Scots at the Battle of Flodden (1513) using "traynes" or long lines of wagons bearing the guns, ammunition and gunners. The guns of the "trayne" were known as "pieces of ordnance", from which comes the word "piece" as applied to modern guns. The guns of the time varied from the Cannon Royal which fired a 70 lb. shot, down to the small Robinet firing a shot of a third of a pound.

The powder charge of the time was roughly half of the shot. The gap between the loose-fitting cannon-balls and the barrel of the gun (known as "windage") caused a great wastage of the explosive force of the powder. For at least three centuries (at least until the end of the eighteenth) the windage was fixed at $\frac{1}{4}$ inch between shot and barrel, no matter how large or small the shot.

Above One of the artillery forts built by Henry VIII in the 16th century on the south coast of England.
Left Early cannon fired crude rockets as well as shot.

Above A gun battery at Sebastopol durin
the Crimean War.
Left A 4·7-inch English gun firing a
Lyddite shell on the Boer position at
Magersfontein. The gun is entrenched i
a pit dug in the open veldt.

Above Inside the Armstrong gun factory at Elswick, England, in 1887.

Improvements went on slowly over the centuries. One was the introduction of the elevating screw. This took the place of the quoin, or wedge, so that the barrel of the gun could be raised or lowered and the range altered in consequence. Other great changes took place during the nineteenth century; William Armstrong and Joseph Whitworth, two engineers who later formed the great armaments firm of Armstrong-Whitworth, introduced many revolutionary ideas. Some of their guns were tried out by the British army during the Boer War (1899–1902); heavy guns had also been used during the Crimean War (1854–56) and during the Franco–Prussian wars from which the military powers of Europe learned a great many lessons.

Meanwhile the navies of the world were building big steel and iron battleships, driven by steam and armed with large guns. But they had little opportunity to test them under war conditions or to study their shortcomings against other maritime powers. The nineteenth century, generally, was a time of peace at sea. Nations were launching battleships which in time became obsolete and went to the scrapyard without having fired a single shot at a real-life enemy.

Left The gun decks of the British warship *Alexandra* (1863).

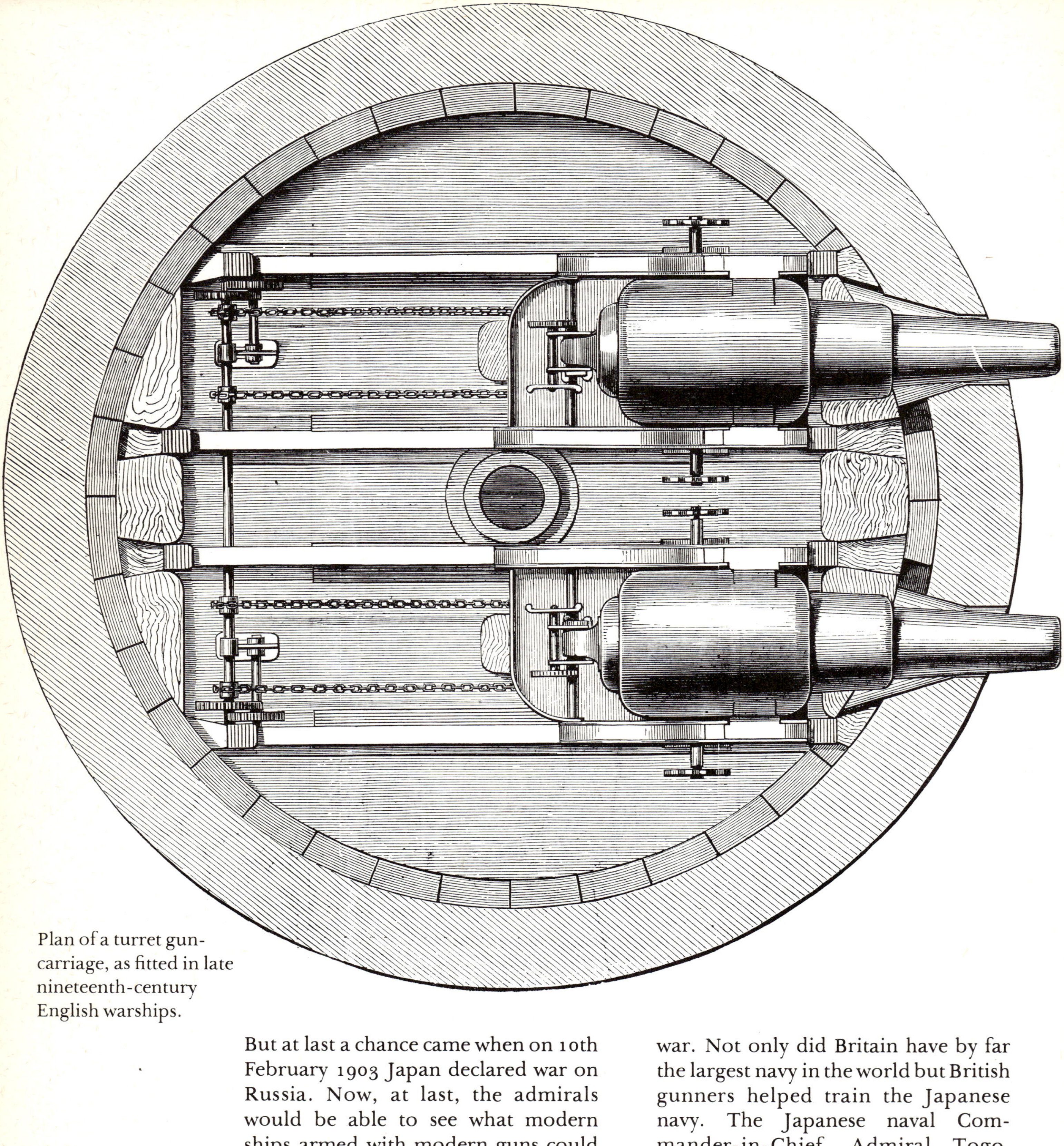

Plan of a turret gun-carriage, as fitted in late nineteenth-century English warships.

But at last a chance came when on 10th February 1903 Japan declared war on Russia. Now, at last, the admirals would be able to see what modern ships armed with modern guns could do. They could test their theories, study the effect of high-explosive shells upon armour, and decide whether the whole science of naval tactics needed revising.

The gunnery experts of the British navy were particularly interested in this war. Not only did Britain have by far the largest navy in the world but British gunners helped train the Japanese navy. The Japanese naval Commander-in-Chief, Admiral Togo, received his early training in H.M.S. *Worcester* on the River Thames, and when he returned to Japan a British captain went with him as adviser.

For more than 400 years the gun had been the principal weapon of naval warfare. The warship was designed

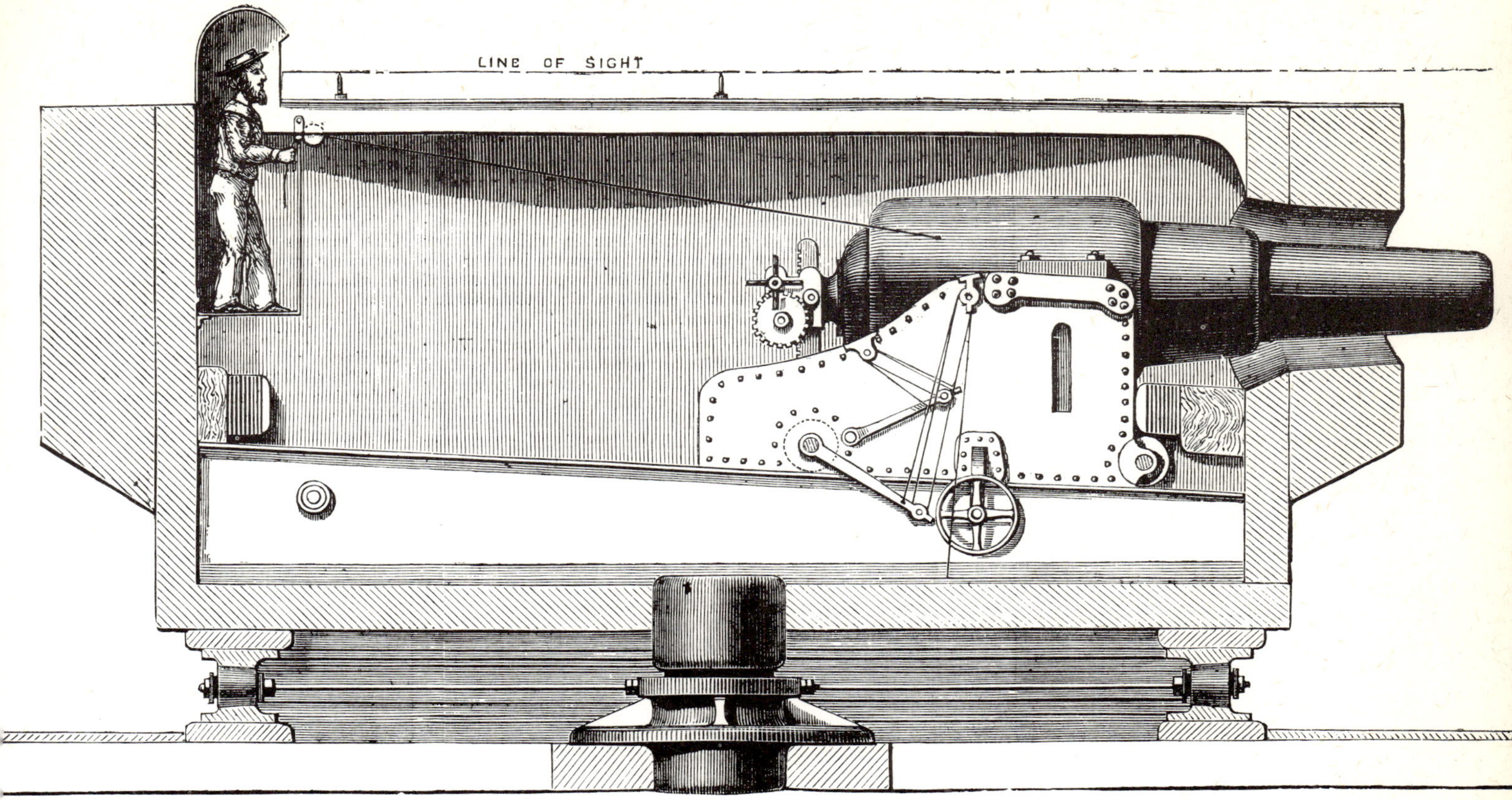

primarily as a gun-carrier and gun-platform. Navies had installed the "breech-loader" and by 1900 all naval guns were "rifled"; that is to say, the barrel contained a number of grooves which passed in a spiral from breech to muzzle. The object of these grooves was to give a fast spin, or rotation, to the shot. This kept it steady during its flight and stopped it "wobbling" or turning over.

The rotation was given by means of a gas-check or a driving-band. The gas-check was a soft metal plate fitted to the base of the shot and slightly larger in diameter than the shot itself. The driving-band was a soft metal band round the lower part of the shot. The force of the explosion drove the gas-check or driving band into the gun and the soft metal was forced into the grooves in the barrel, giving it the necessary twist.

The projectiles used in the heavy guns of this time were nearly all the same shape, elongated with conical heads. Some were made with a blunt head, as this was thought to give a better penetration through armour. The shot varied from the small three-pound shell of the quick-firing gun to the monster shell weighing 2,000 pounds (900 kg) and standing, with its charge, some nine feet high.

Above Section of the gun turret shown on page 10.

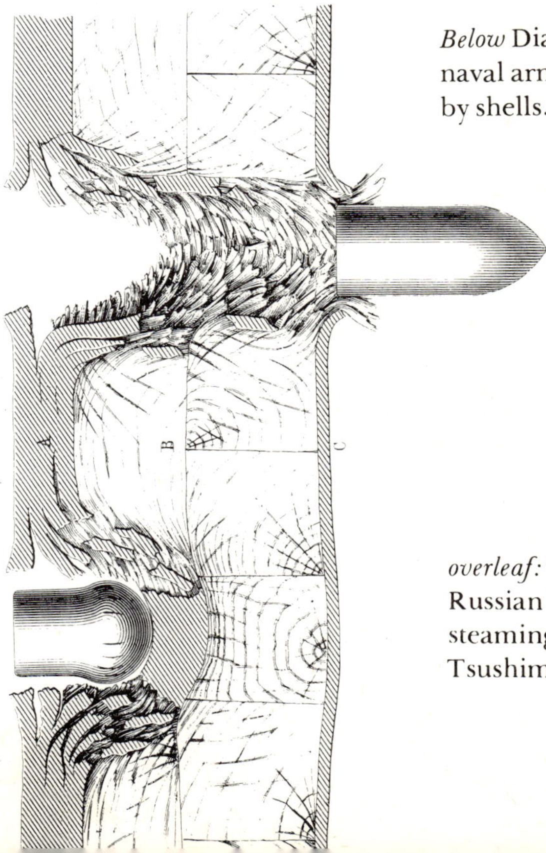

Below Diagram showing naval armour pierced by shells.

overleaf: The Russian fleet steaming towards Tsushima.

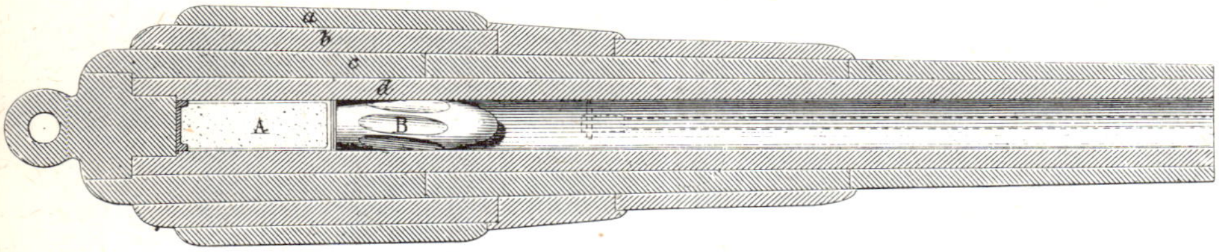

Cross-section of a Whitworth gun and (*right*) its rifled shot.

Shells also varied in construction. There was solid shot made of hardened iron or steel, and used to pierce hardened armour-plates; common shell which carried a large bursting charge but which was also strong enough to pierce armour-plate of moderate thickness and to explode inside the ship; the shrapnel shell with thin walls, filled with bullets and a charge of powder at the base. This did not pierce armour but had a deadly effect upon men exposed in the open.

Both Russia and Japan had modern warships, equipped with guns ranging from the powerful 12-inch primary armament down to the (still large) 6-inch. The advantage, however, lay on the side of the Japanese whose crews were better trained and had a fighting spirit that was completely missing from the Russians.

Below The Japanese warship Idzumo and its armament.

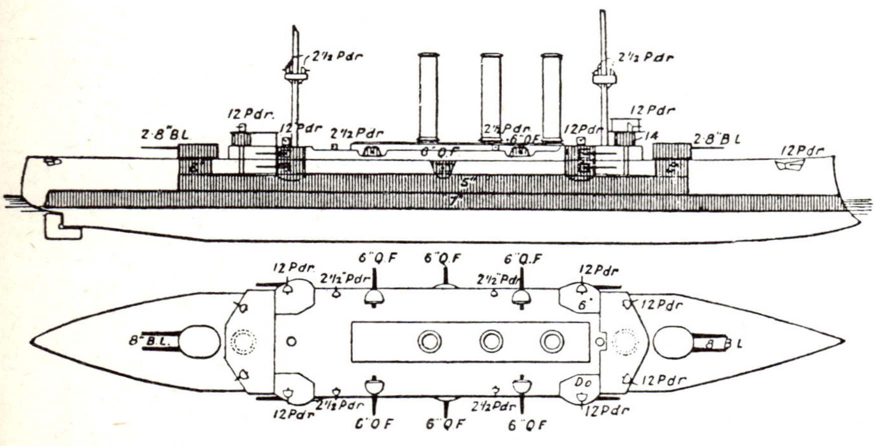

Length, 400—431 ft.; 9436—9850 tons; Speed, 20—22 knots; Completed, 1901; Armament, 4—8 in., 14—6 in., 12—12 pr., 8 small.
*12—6 in. guns.

So it was, with the eyes of the world focussed on his fleet, that Vice-Admiral Rojesvensky, flying his flag in the *Suvoroff,* led out Russia's Baltic Fleet during October 1904. The ships began a long haul that was to take them three-quarters of the way around the world in order to reach the Pacific and, hopefully, destroy the Japanese fleet under Admiral Togo.

On May 25th, 1905, sea-stained and weary, the fleet left Shanghai on the last stage of its incredible 18,000-mile journey . . . and Admiral Togo was waiting for it. Togo had been kept fully informed of its progress and he knew that Rojesvensky was heading for the base at Vladivostock. Togo's only problem was to know which of the three routes the Russian fleet intended to take. But he guessed rightly and the first contact was made at dawn on May 27th when a Japanese auxiliary cruiser, *Sinano Maru,* almost ran down one of the Russian hospital ships steaming in the rear of his fleet. Not long afterwards the cruiser sighted the rest, three armoured squadrons, a cruiser squadron, a scout division and two cruisers that were co-operating with nine destroyers.

This information was wirelessed back to Admiral Togo. Soon afterwards, when the large cruiser *Idzumo* appeared to shadow the Russian ships, Rojesvensky immediately ordered his fleet to steam at full speed.

Right: Map showing position of opposing forces, May 27, 1905. *Below* Admiral Togo directing operations of the Japanese fleet from the bridge on his flagship *Mikasa*.

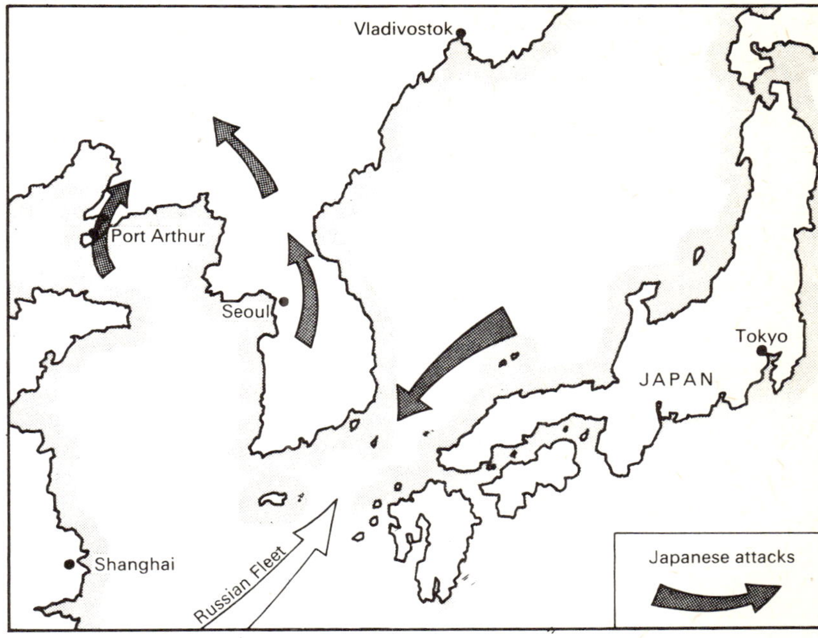

A strong wind soon shredded away the morning mist, causing the sea to rise so that the Russian ships, especially the destroyers and torpedo boats, were burying their noses in the grey-green sea and swinging wildly from side to side.

At midday, after every officer and seaman aboard the Russian fleet had drunk the health of the Tsar—the day was the anniversary of his coronation—everyone went to their action stations. Just before 1.30 p.m. the grim lines of Togo's ships were sighted some six miles away to the north-east. At the masthead of his flagship *Mikasa* streamed a signal which echoed that of Nelson, flown at Trafalgar a hundred years before: "The rise or fall of the Empire depends on today's battle. Let every man do his utmost."

The two fleets were approaching each other in "line-ahead" formation when at 1.45 p.m. Togo made a bad mistake. Against any other enemy it would have been fatal. He ordered his battleships to turn "in succession" together across the bows of the oncoming Russian fleet. When a number of ships turn "in succession" when steaming in single-line ahead, the turning point remains a fixed target for several minutes. The opponent has only to range his guns on this one point to hit each ship as it reached it.

Rojesvensky now had a wonderful opportunity. His four leading ships

could have opened a devastating fire from sixteen 12-inch guns and rapidly changed the outcome of the battle. But for some strange reason, although the long barrels of his guns were pointed straight at the enemy, they did not fire. Togo was able to carry out the first part of his manoeuvre without a single shot being fired. When the range was only 6,400 yards, the *Suvoroff* at last began firing, yet the gunnery and shells were so poor that the opportunity was completely lost.

Below Togo's flagship the *Mikasa*.

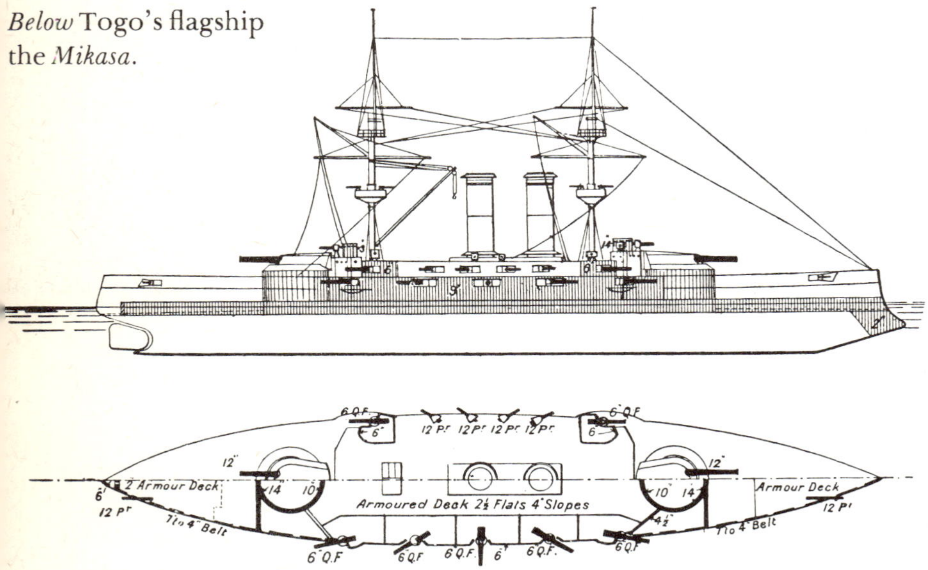

Length, 400 ft. ; 15,200 tons ; Speed, 18·5 knots ; Completed, 1902 ; Armament, 4—12 in., 4—10 in., 10—6 in., 20—12 pr., 20 small.

Below The Japanese armoured cruiser *Azama*.

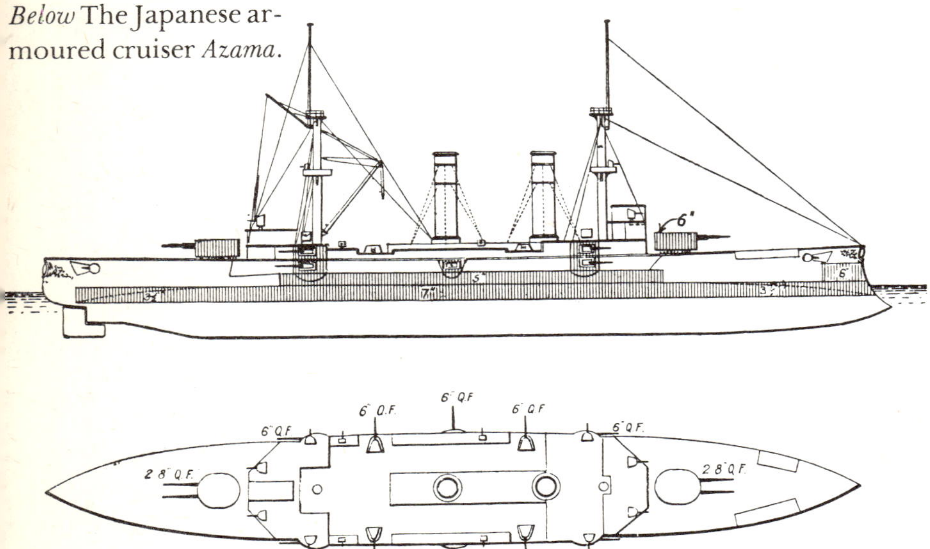

Length, 408 ft. ; 9700 tons ; Speed, 22·1—23 knots ; Completed, 1899 ; Armament, 4—8 in., 14—6 in., 12—12 pr., 8 small.

The Japanese ships also opened fire and soon got the range. The better-trained Japanese gunners, firing shells with a bursting charge that was seven times more powerful than that of their enemy, concentrated on the *Suvoroff,* smashing her time and time again. Her guns were hurled from their mountings and her steel plates crumpled like tinfoil until she finally fell out of line and wallowed astern, battered and burning.

The Japanese then shifted their target to the *Oslyababa,* their shells smashing into her with great gouts of red flame until, ablaze from stem to stern, she began to sink and finally disappeared—the first modern battleship to be sunk by enemy gunfire.

A general engagement then took place at fairly close range but the Japanese rate of fire was so fast and so accurate that the Russian ships had hardly a chance. As a Japanese officer said later, "After the first twenty minutes the Russians seemed suddenly to go to pieces, and their shooting became wild and almost harmless." At 4 p.m. fire broke out on the *Sisoy Veliky* and she fell out of line, her place being taken by the *Alexander III* whilst the *Navarin* also fell out of the line and others came in to fill the vacant space.

It was no longer a battle, it was a massacre. The Russian ships were singled out one by one until at 5 p.m. they were crowded together. Togo's battleships were still pouring in a relentless hail of fire from the east and a lighter but still devastating fire came from his cruisers to the north. During the whole of the battle a British naval adviser, Captain Pakenham R.N., had been on board Togo's *Mikasa.* This officer, tall and immaculately dressed with a monocle in his eye, sat calmly in a deck chair whilst the flagship's great guns roared out all about him, calmly taking notes of everything that was happening.

Right Destruction of the *Borodino*.

The fateful day was drawing to a close, a brilliant red sky in the east marking the setting sun. Admiral Nebogatoff, who had taken over command in the ageing *Nicholas I*, signalled "follow me" in the desperate hope of making off to the south-west with five ships of the scattered fleet. At daybreak, however, Togo's battle fleet, still intact, steamed across his bows. With hardly any hesitation, Nebogatoff hoisted the signal of surrender.

Later that day a count revealed what a disaster Tsushima had been for Russia. Of her eight battleships, six had been sunk and the others captured. A further twenty-two ships had been sunk or had gone aground; six had surrendered. Only six, somehow, had managed to reach a neutral port where they were interned. Only three ships out of the whole fleet managed to reach the safety of Vladivostock.

By comparison the Japanese had only three torpedo boats sunk and a number of small craft and one armoured cruiser, the *Azama,* damaged.

The ships of the two main fleets, with their guns, were as shown in the table opposite.

Tsushima taught the other maritime nations of the world a great many lessons. Britain was the best informed, for she was not only an ally of Japan but had the first-hand reports of her observer, the unflappable Captain Pakenham. One great lesson, however, seems to have gone unnoticed. For the first time in history an eastern power had shown that she could produce a fighting fleet that was equal to almost any other then afloat.

In the relatively short space of thirty-five years Japan was to increase her naval strength until she was capable of challenging one of the world's greatest naval powers . . . the United States of America.

RUSSIAN	*Vessel and guns*	12″	10″	9″	8″	6″	smaller
IRONCLADS	*Suvoroff*	4	–	–	–	12	42
	Alexander III	4	–	–	–	12	42
	Borodino	4	–	–	–	12	42
	Oryol	4	–	–	–	12	42
	Oslyaba	–	4	–	–	11	46
	Navarin	4	–	–	–	8	32
	Nicholas I	2	–	4	–	8	20
	Admiral Nakhimoff	–	–	–	8	10	18
FIRST-CLASS CRUISERS	*Oleg*	–	–	–	–	12	22
	Aurora	–	–	–	–	8	32
	Dimitri Donskoy	–	–	–	–	6	38
	Vladimir Monomakh	–	–	–	–	6	32

JAPANESE	*Vessel and guns*	12″	10″	9″	8″	6″	smaller
IRONCLADS	*Mikasa*	4	–	–	–	14	32
	Shikishima	4	–	–	–	14	32
	Fuji	4	–	–	–	10	14
	Asahi	4	–	–	–	14	32
LARGE ARMOURED CRUISERS	*Kasuga*	–	1	–	2	14	16
	Nisin	–	–	–	4	14	16
	Idzumo	–	–	–	4	14	19
	Yakumo	–	–	–	4	12	19
	Asama	–	–	–	4	14	19
	Adzuma	–	–	–	4	12	19
	Tokiva	–	–	–	4	14	19
	Ivate	–	–	–	4	14	19

Chapter Two

Liège 1914

"The Whole Earth Shook"

The fighting between the Japanese and Russians was even fiercer on land than on sea. Russia held the strongly fortified harbour of Port Arthur; Japan was determined to wrest it from her. The siege of the harbour fortress began on July 2nd, 1904, and the bitter fighting lasted with hardly a pause for another six months.

At first the ships of the Japanese navy lay off the harbour and bombarded the forts with their huge 12-inch guns, but seeing that progress was slow they landed some of the guns, with seamen-gunners to serve them, mounting them in two of the outlying forts that they had previously captured by direct assault. In October a heavy bombardment by the large siege and naval guns went on for four days without a pause, inflicting tremendous damage.

The use of monster guns for siege purposes was nothing new. Indeed, such large and cumbersome but very effective pieces of ordnance had existed from the Middle Ages. When the English evacuated Mont St. Michel in 1424 one of the guns they left behind measured 12 feet in length and had a calibre of 18 inches. It weighed $5\frac{1}{2}$ tons. It can still be seen today. The most famous of all heavy siege guns are the monsters used by Sultan Mahomet II when he steadily demolished the walls of Christian Constantinople in 1453. One of them was 17 feet long, a calibre of 25 inches and, weighing nearly 19 tons, required a team of 60 oxen and 30 wagons to move it. 200 men had to march beside it to support its weight whilst another 250 went ahead to smooth the roads. The gun fired a stone cannon ball weighing more than 600 lbs, and had a range of about a mile.

Another famous large calibre gun of the period can still be seen at Edinburgh Castle. Known as "Mons Meg" it is believed to have been made at Mons in Belgium, then a large foundry centre, in about 1460. It is over 13 feet long, has a calibre of $19\frac{1}{2}$ inches and fired a granite ball of 330 lbs. It also had an astounding range of 2,876 yards (2,630 metres).

What is meant by "calibre"? In the early days of cannon, shot was circular and made of stone and the gun was

Below Giant bronze cannon used at the siege of Constantinople in 1453.

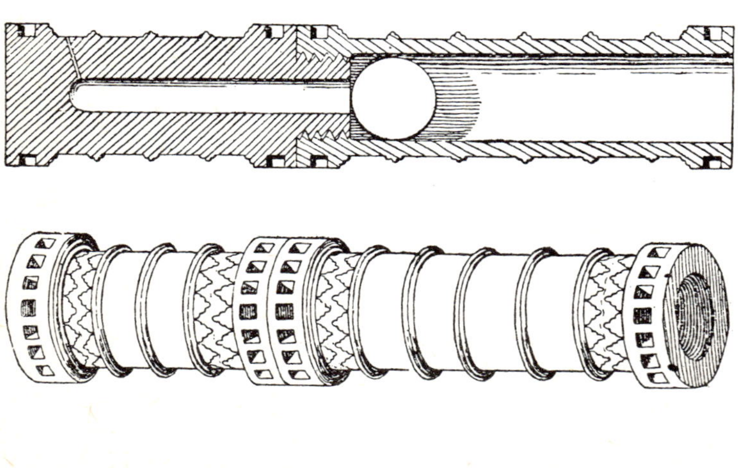

Left Mons Meg at Edinburgh Castle.

known by the weight of shot it fired. This was also true when iron balls were introduced. In fact it then became easier, for the density of iron remained constant and it was simpler to estimate. A cast-iron shot weighing 6lb had a diameter of 3·55 inches, although it was customary to make the "bore" of the gun a little larger to take the shot. (The "bore" is the open space inside the barrel along which the shot travels.) Thus a 6-pounder was a gun with a bore of about 3·67 inches and the diameter of this bore became the calibre of the gun, a 12-pounder having both bore and calibre of 4·62 inches, an 18-pounder 5·3 inches, and so on.

When circular balls went out of fashion and the more familiar elongated "shells" came in, this system had to be altered. A 12-inch cast-iron shot weighed about 230 lbs, but the solid, elongated shell fired from the same gun weighed about 700 lbs. To describe the gun as a "700-pounder" would have been misleading for had the same shot been spherical it would have needed a bore of more than 17 inches. So finally, and with only a few exceptions, guns were known by the *calibre*—in inches or millimetres.

In the time of the Tudors, however, guns were given names which applied to their size and some of these names are quite interesting. The range of guns used during the sixteenth century was:

NAME OF PIECE	CALIBRE	WEIGHT OF SHOT	CHARGE OF PROPELLANT POWDER
Cannon Royal	8·5 inches	74 lbs	30 lbs
Cannon	8·0 inches	60 lbs	27 lbs
Bastard cannon	7·0 inches	42 lbs	20 lbs
Demicannon	6·4 inches	32 lbs	18 lbs
Petro	6·0 inches	26 lbs	14 lbs
Culverin	5·2 inches	18 lbs	12 lbs
Basilisk	5·0 inches	14 lbs	9 lbs
Demiculverin	4·0 inches	8 lbs	6 lbs
Saker	3·65 inches	6 lbs	4 lbs
Minion	3·5 inches	5·2 lbs	3 lbs
Falcon	2·5 inches	2 lbs	1·2 lbs
Falconet	2·0 inches	1 lbs	0·4 lbs
Serpentine	1·5 inches	0·5 lbs	0·3 lbs
Robinet	1·0 inches	0·3 lbs	0·18 lbs

A number of these guns were carried by warships and in time the names were to be changed to the weight of the cannon balls they fired. Thus: the Bastard Cannon became the 42-pounder, the largest gun carried in ships of war during the time of Nelson; the Demicannon the 32-pounder; the Petro the 24-pounder; the Culverin the 18-pounder; the Basilisk the 12-pounder; the Demiculverin the 9-pounder and the Saker the 6-pounder.

Such guns, however, which had barely changed down the centuries, became obsolete when shells and rifling came into general use.

The large land guns were very difficult to move from place to place. The cannon royal, for example, weighed 8,000 lbs. It needed fifteen pairs of horses to pull it, with a further horse known as the "thriller" trudging behind to act as a brake when going downhill. Even a small gun such as a falconet needed two pairs of horses and a thriller.

The difference between siege and field guns was mainly seen in their weight. If a gun could be hauled along rough roads or tracks of the time and hurried into battle, it was a "field" gun. The culverin was the largest of these, though rarely used. The siege gun, on the other hand, was pulled and pushed into position where it remained, pounding away at its target until the fortress was taken or the siege was lifted. Almost all siege guns were cannon—the cannon royal, cannon and demicannon, with calibres ranging from 8 to 6 inches. Oxen were usually used to haul these monster guns from place to place.

Nevertheless, all the big guns of the past seemed to become insignificant beside the huge weapons that the next major conflict was to produce. This was the Great War of 1914–18, destined to be truly a World War, for in time it was to involve nearly every country in the world. Hostilities began on August 1st, 1914, between Germany and France. Germany immediately put into action a plan of attack known as the Schlieffen Plan.

Part of this plan called for two German armies to invade Belgium—then still neutral—across the River Meuse and to encircle the city of Liège. Then, having captured the city and using the element of surprise, they would link up with each other and sweep towards the Channel coast.

The only question was, how strong was Liège itself? Thirty years before, a chain of twelve forts had been built in a circle around the city and on either side of the River Meuse to prevent, or at least deter, such an invasion.

Belgium, ill-prepared for war, had allowed the forts to fall into some disrepair. In any case, their 7-foot thick concrete walls were only designed to withstand bombardment of guns of up to 21-cm calibre. The German guns that arrived to "crack this tough nut" were truly enormous—their 42-cm howitzer was the largest gun yet to be fired in battle. Made by the famous firm of Krupp it was known as "Big Bertha", or, more accurately, "Fat Bertha". It had a crew of 200, weighed 75 tons and was capable of firing ten rounds an hour, projecting a 2,200 lb shell more than 15,000 yards (13,716 metres or 9 miles). It had to be transported by rail in five sections. A smaller but equally effective gun used by the Germans was the Austrian-built Skoda, a 30·5-cm howitzer which weighed 28 tons and fired a projectile weighing 846 lbs over 13,000 yards (11,887 metres, or more than 7 miles). Its great advantage was that it was transportable by road and could be assembled and made ready to fire very quickly.

Against these monsters the Belgians had only guns of medium calibre, the largest being 21-cm. It was a one-sided affair from the start.

The defence of Liège was in the hands of General Gerard Leman; the invasion force was commanded by Erich von Ludendorff who was to become one of the best commanders of the war.

The night of August 5/6th was dull and overcast. There was thunder in the air, and occasional flashes of lightning that were soon to pale before the opening fire of the Belgian guns in Fort Fleron. The following day General Leman sent some of his forces to safety deep within Belgium and ordered his fortress gar-

risons to hold out to the end. He wanted to buy precious time so that Liège might not be captured. Within hours of this order, however, strong groups of German infantry had passed between the forts and penetrated into the city. The raw and inexperienced defenders were only too ready to surrender to the men in field grey uniform who had so suddenly and so alarmingly appeared.

With Liège in German hands, only the forts remained. More German troops were brought up from the reserve and took up their siege positions. On the 8th, two light howitzer batteries (there were six guns to a German battery) opened fire, and the first shells struck the crowded parapets of Fort Barchon killing more than 30 men outright. This was enough for the garrison. It surrendered soon afterwards.

By the 11th the same batteries had pounded the neighbouring Fort Evegnée into submission, then the two "Big Berthas" were brought up. A German eye-witness spoke of the tremendous roar that burst from the first of the two to be fired. As it did so there was a cheer from the gun's crew, a cheer which "accompanied the shell as it howled and snorted along the high trajectory to its target . . . sixty seconds ticked by—the time needed for the shell to traverse its 4,340-yard-high trajectory—and everyone listened in to the telephone report of our battery commander, who had his observation post 1,625 yards from the bombarded fort, and could watch at close range the column of smoke, earth and fire that climbed to the heavens."

A delayed-action fuse made sure that the 2,000-lb shell penetrated within its target before exploding. It was the largest shell ever to be fired in action. The sound caused by the explosion was so loud that Belgians in other forts believed that a direct hit had been scored on the fort's magazine. As other shells followed the first, however,

Above The 75-ton "Big Bertha" built by Krupps.

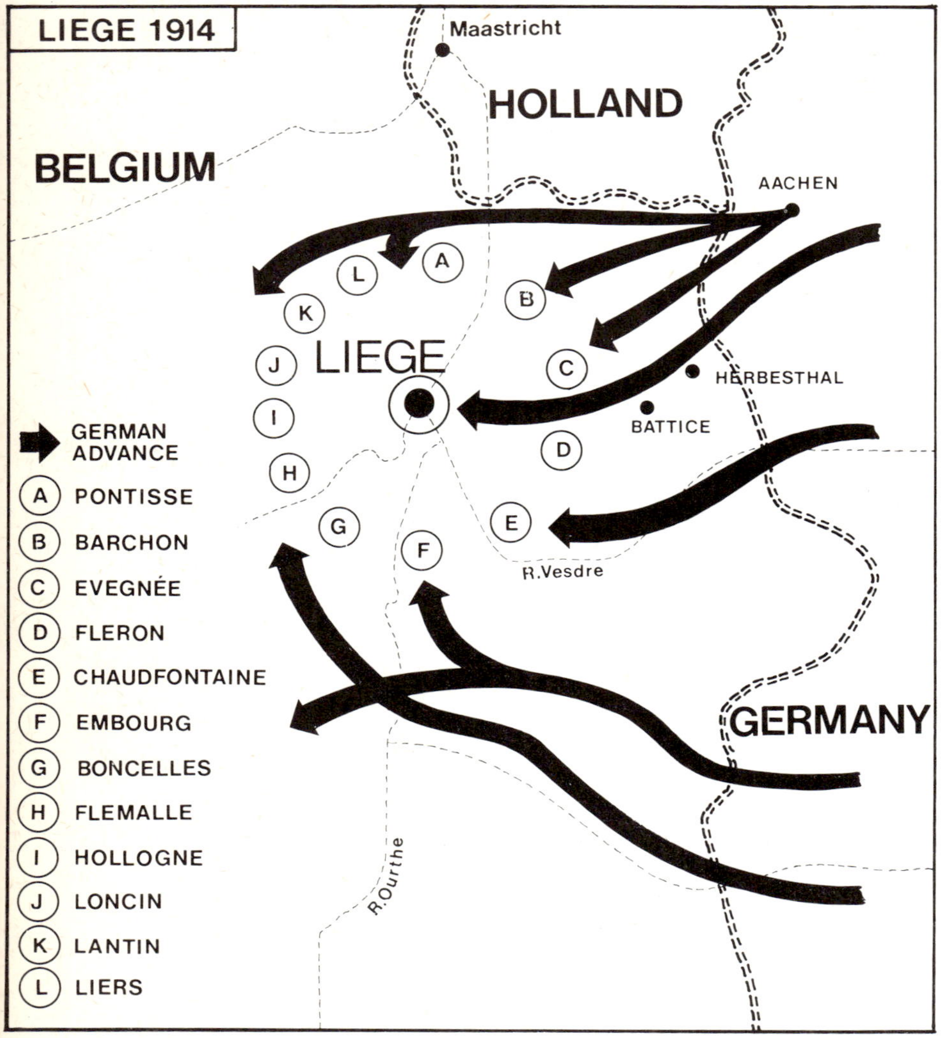

the unfortunate defenders realized that they were faced with something quite new in land warfare—a huge gun with an almost unbelievable impact. The range had to be corrected after each explosion and the last shell of the day—the eighth to whine across nine miles of Belgian soil—actually landed on the fort. There was an ear-splitting explosion and horrified onlookers saw large slabs of concrete, jagged pieces of armour plating, steel gun cupolas and even bodies, being hurled skywards in a welter of flame and yellow-brown smoke.

Soon after noon Fort Pontisse, the first target of the "Big Berthas" surrendered, and the great guns trundled on to the next. Five hours later Fort Embourg surrendered, too. On the next day, August 14th, two more, Liers and Fleron and on the 15th Chaudfontaine, Boncelles and Lantin also gave in.

Now only the forts of Loncin, Hollogne and Flemalle still continued to fight. The first was the key to the whole system, for General Leman had established his command post within

Below A big gun at the finishing lathe.

Above The destruction at Fort Loncin.

the stronghold. On the 15th the "Big Berthas" moved into position and concentrated their fire on Fort Loncin. Leman was later to describe the shelling: "We heard the rushing of air which increased in intensity until it became a furious hurricane roar and ended in a dreadful crash of thunder, fountains of earth and smoke were thrown into the air, and the whole earth shook."

Although shocked by the force of the explosions and blinded by the swirling, throat-rasping smoke and fumes, the garrison bravely fought back. One hour passed, then two, but as the dreadful minutes ticked away there came an explosion which even surpassed what had gone before. One of the great shells had penetrated deep inside the fort, its delayed-action fuse had gone off and the main magazine had exploded, ripping and tearing the solid concrete as if it had been paper.

That was enough. The tricolour flag fluttered down in surrender. Groups of German pioneers moved forward to give assistance to the garrison—or what was left of it. They stumbled forward over piles of twisted steel and pulverized rubble and managed with great difficulty to drive a path through the wilderness of shapeless debris, all that remained of a once-proud fort. One of the heavy cupolas had been blown from its foundations and had fallen upon its dome looking for all the world "like a monstrous tortoise, lying in its shell."

The ghastly and abrupt end of Fort Loncin was enough for the surviving forts. They hurriedly surrendered before the two monster guns could point their grim barrels in their direction. Now, with the forts no longer a threat and the city of Liège in their hands, the road was open for the Germans to move forward into Belgium and, hopefully, to the Channel ports and even Paris itself.

But the Belgian resistance had delayed the German onslaught long enough for the French to reorganize their forces to face an attack from this unexpected quarter. It also allowed Britain time to send across her Expeditionary Force of 100,000 men who were soon to earn eternal glory for themselves as "The Old Contemptibles".

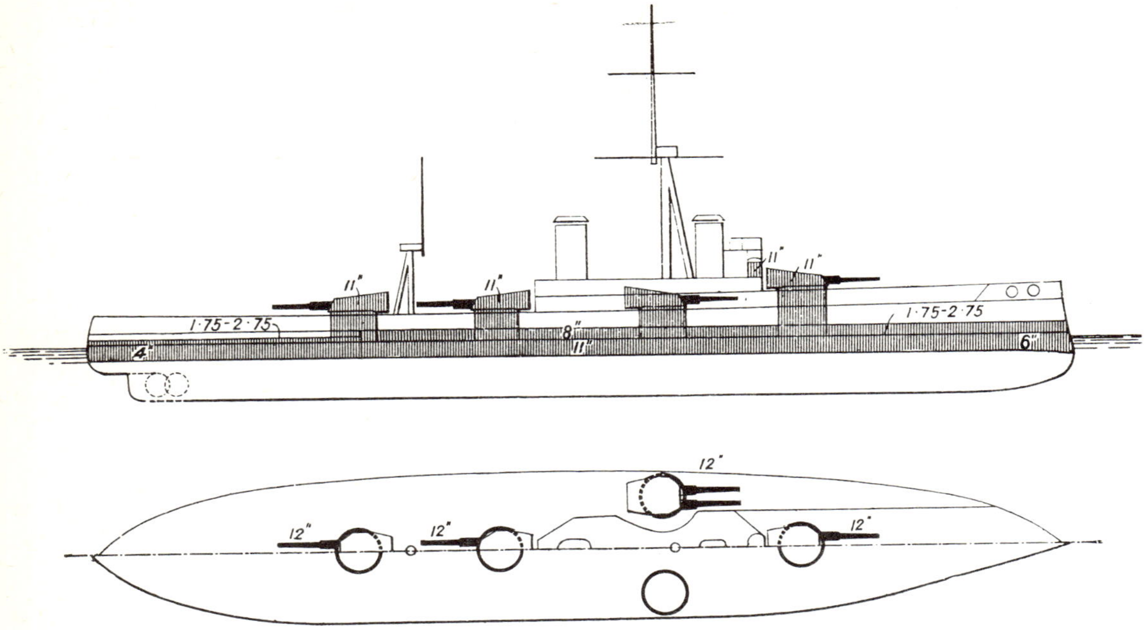

Dreadnought.—Length, 490 ft. ; 17,900 tons ; Speed, 21·8 knots ; Completed, 1906 ; Armament, 10—12 in., 24—12 pr.

Below Dreadnought.

Chapter Three

Coronel and Falkland 1914

Old Guns, New Guns

In 1905, the year of Tsushima, the leading naval powers of the world possessed 120 capital ships. Of these Great Britain had one third, but her position as the world's greatest navy was already threatened. The United States, France, Italy, Japan and especially Germany, were also building bigger and more heavily armed warships. As if that was not enough, British politicians were demanding that their country should limit the number of warships being built.

But Britain had a man who had the answer. He was John (later Lord) Fisher, who had dreamed of a new type of warship, one that would make all others out-of-date. Fisher had also learned a great deal from the reports of Tsushima, which enabled him to produce his "secret weapon". This was the revolutionary warship *Dreadnought,* completed in the record time of a year and a day; the first keel plate was laid at Portsmouth on October 2nd, 1905. She was launched on February 10th, 1906, and steamed out on her trials on October 3rd, 1906.

Dreadnought carried ten 12-inch guns set in pairs in five turrets—three on her centre line and one on each side between her funnels. In this way she could fire six guns forward, or aft, and a broadside of eight. Apart from a few small guns to counter attacks by torpedo boats, she carried none of the guns of medium calibre known as the "secondary armament" carried by all others at this time.

As far as could be seen, her one defect was the position of her mast, which was stepped abaft the fore funnel. This meant that the control of the guns was hampered as the gun-control platform was frequently smothered in smoke. This was corrected in later ships.

Dreadnought was fitted with turbine engines and steamed at 21·2 knots, displacing 17,900 tons of water. Her guns, armour and speed set the standard for a race in armaments. Two fine British battleships, the *Lord Nelson* and the *Agamemnon* then building, were completed in June 1907 and December 1908, yet the *Dreadnought* made them obsolete before they were even built.

In the days before *Dreadnought,* most battleships had a complexity of armaments, with four 12-inch guns in two turrets, one forward and one aft of a central structure known as the citadel. The citadel was packed with smaller guns (usually 6-inch) known as the secondary armament. On top of the citadel were the smaller guns (12 and 3-pounders) for repelling attacks from destroyers and torpedo boats.

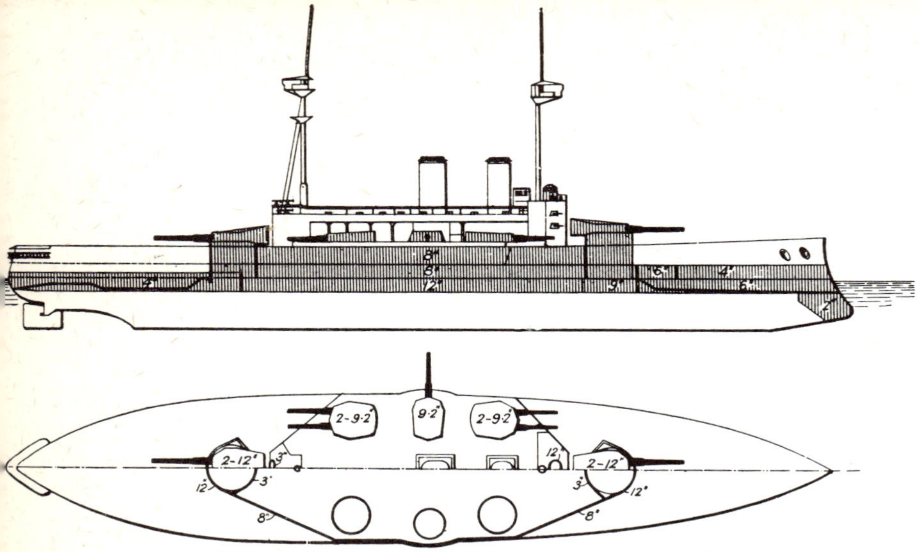

Length, 410 ft. ; 16,500 tons ; Speed, 18·75–18·9 knots ; Completed, 1908 ;
Armament, 4—12 in., 10—9·2 in., 24—12 pr., 5 small.

Top and centre Agammemnon; *bottom* Nelson.

The firepower of one of these older battleships was nothing compared to the concentrated fire of the *Dreadnought,* capable of firing eight out of her ten 12-inch guns in one mighty broadside.

The rest of the great maritime powers gasped, took stock of the situation, then hastily began to build dreadnoughts of their own. From the day the *Dreadnought* was launched, every other battleship in the world was to be classed as dreadnought, pre-dreadnought or super-dreadnought. This distinction referred both to the battleships and battle-cruisers. The former were the direct descendants of the old line-of-battle ships of Nelson's time, and formed the powerful, medium-speed backbone of the navy. There were originally seven dreadnoughts, including the type ship, which differed from each other only in minor details. Because of the effect on her gunnery control, all other dreadnoughts had two masts.

The next dreadnoughts to be built, the *Neptune, Hercules* and *Colossus*, were slightly larger, and their 12-inch guns were fitted with longer barrels. These guns were not successful, however. They were believed to have a tendency to "droop" in the middle. The layout of the guns was improved, all ten guns being able to fire in one broadside. This was done by having a superimposed turret aft and having the two other turrets *en echelon* on either side of the ship.

The next four dreadnoughts, of the *Orion* class, were fitted with larger guns, 13·5-inch, and were greatly superior to their predecessors. All guns could fire on either broadside and the secondary armament of 4-inch guns was centralized, a practice that was adopted in all later battleships.

A later class still, the Queen Elizabeth class, mounted even larger guns—eight 15-inch. They were also the first battleships to have oil-fired boilers which gave them a speed of 25 knots, almost that of a battle-cruiser. The largest guns carried in a British battleship were the nine 16-inch guns of the *Nelson* and *Rodney*. These were set in threes in three turrets set forward of the main control section, and only smaller guns were aft. This, presumably, was to suggest that both ships would never present their stern to the enemy and run away.

The battle-cruisers came into the dreadnought class. They were also the creation of Lord Fisher. He wanted a warship which would combine the speed of a fast cruiser with the striking power of a powerful battleship. To provide this speed, however, she had to be less heavily armoured than a battleship which with her better protection, could destroy a battle-cruiser in a straight fight. A battle-cruiser, however, could use her faster speed to escape. Just as the all-big-gun battleship of the dreadnought class made all other existing capital ships obsolete, so Fisher's first three battle-cruisers, *Inflexible, Invincible* and *Indomitable,* outclassed all the big armoured cruisers of other nations.

Most of these suddenly out-dated cruisers carried two or four 9·2 or 8-inch guns and had a top speed of 21 knots. By comparison, Fisher's new battle-cruisers had an armament of eight 12-inch and sixteen 4-inch guns, and a speed of more than 25 knots. They obviously outclassed all other heavily armoured cruisers of the time and presumably, although they had not yet been tested, could demolish anything afloat except the slower but more heavily armoured battleships.

These new battle-cruisers, together with all dreadnoughts of the time, were fitted with the huge breech-loading guns, rifled, and firing cylindrical and explosive shells over vast distances. Despite the new threat of torpedoes

Orion.

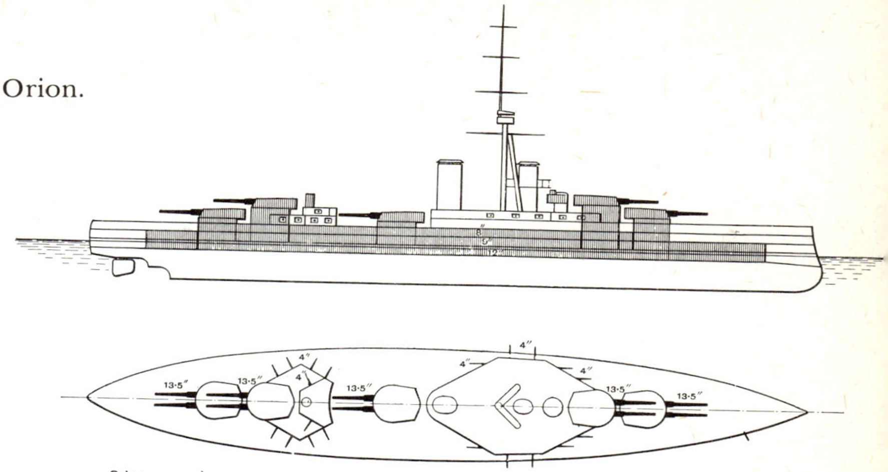

Orion, Conqueror, Monarch, Thunderer: Length, 545 ft. ; 22,500 tons ; Speed, 21 knots ; Building ; Armament, 10—13·5 in., 16—4 in. ; Completed, 1911-12.

Ajax, Audacious, Centurion, King George V.: Length, 555 ft. ; 23,000 tons ; Speed, 21 knots ; Building ; Armament, 10—13·5 in., 16—4 in. ; Building.

Above Orion.

Above Neptune.

and bombs from aircraft, they were still *the* offensive weapon in naval warfare.

Less than eight years after the *Dreadnought* had steamed on her trials, Admiral Fisher's ideas were to be put to the test. After Germany violated Belgian neutrality Britain issued an ultimatum and when this expired on August 4th, 1914, World War One began.

The war opened well for Britain. On August 28th, 1914, German and British forces clashed in Heligoland Bight. Here a battle-cruiser squadron sank three German cruisers, the *Mainz, Ariadne* and *Köln* and a destroyer, with no loss to themselves.

News of this victory, however, was soon to be followed by further news —this time of a crushing defeat.

The only German squadron operating at sea (the rest were blockaded in harbour) was composed of five warships commanded by Admiral Graf von Spee. He had two heavy cruisers, the *Scharnhorst* and *Gneisenau* and three light cruisers, the *Leipzig, Nürnberg* and *Dresden.* A small force of British warships was near the last reported position of this squadron and the Admiralty ordered its commander, Admiral Cradock, to "destroy the cruisers".

Towards the end of October, Cradock was steaming along the Chilean coast when he made contact with the enemy. He was flying his flag in an ageing armoured cruiser called the *Good Hope*, and a smaller armoured cruiser, the *Monmouth.* His third ship was the light cruiser *Glasgow* and the fourth, an armed liner, *Otranto*.

Cradock was doomed from the outset. His flagship, the *Good Hope,* had been built in 1902 and was armed with two old-type 9·2 guns and sixteen 6-inch. She was weakly armed for her 14,000 tons and presented an immense target to the enemy. The 10,000-ton *Monmouth* had been built about the same time as the flagship and was even more weakly armed. Her main armament consisted of fourteen 6-inch guns. The *Glasgow* was a light cruiser which carried only two 6-inch guns whilst the *Otranto,* even more defenceless, had been ordered to keep away by Cradock; with her high ocean-liner sides and inferior guns she was too vunerable to be pitted against warships.

Below The German heavy cruiser *Gneisenau.*

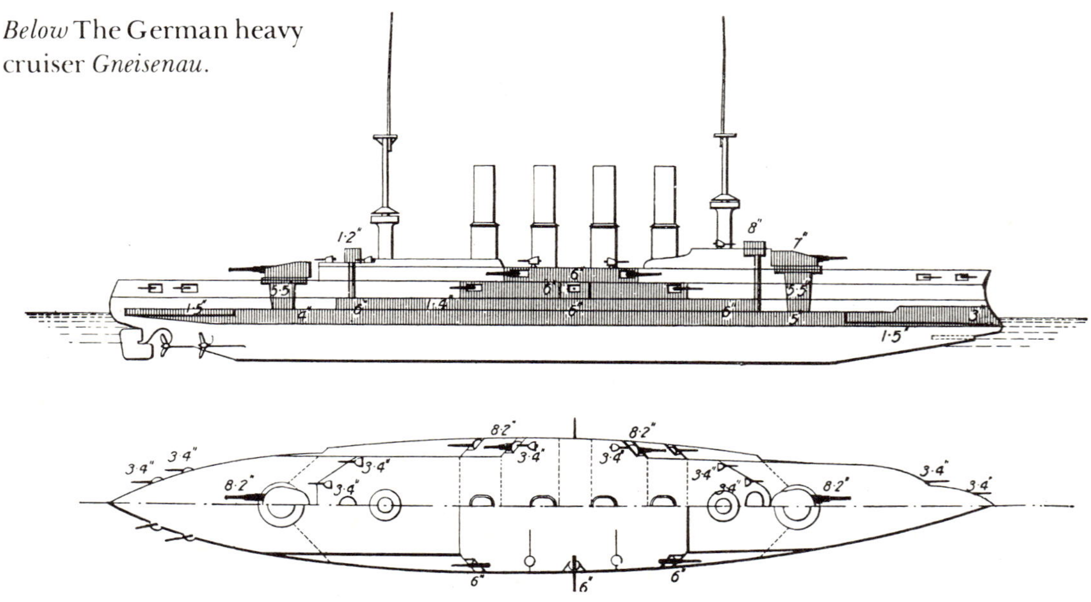

Length, 450 ft. ; 11,420 tons ; Speed, 22·5—23·8 knots ; Completed, 1908 ; Armament, 8—8·2 in., 6—6 in., 20—3·4 in., 14 small.

Contact was made during the afternoon of November 1st. Both squadrons were steaming south in a strong wind and considerable sea. Graf von Spee cleverly avoided action until dusk when, with the British ships silhouetted against the setting sun, he would have a great advantage.

The *Scharnhorst* and *Gneisenau* were classed as heavy armoured cruisers, 11,000-ton sister ships, armed with eight new 8-inch and six 6-inch guns each. The *Dresden, Nürnberg* and *Leipzig* were light cruisers of 3,300 tons, each mounting ten 4·1 guns. Given even reasonable shooting the end was a foregone conclusion, especially as the two larger German warships were the crack gunnery ships of the German navy. The weight of the *Good Hope*'s armament and that of her two principal opponents was, on paper, not so very different, but they were two to one and the broadside of the *Monmouth* was only 900 lbs. Thus the apparent weight of the metal thrown by von Spee without his light cruisers was 3,906 lbs, whilst Cradock could only discharge 2,460. But even this was misleading: owing to the heavy sea that was running the *Good Hope*'s lower 6-inch guns (mounted in barbettes, in pairs, one above the other) could not be used.

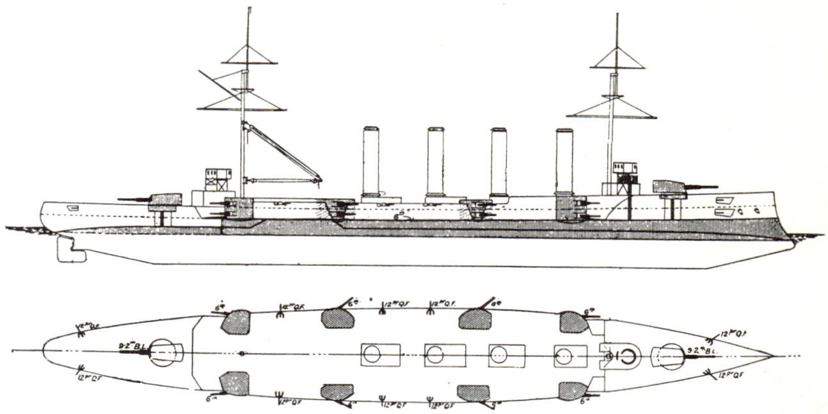

Length, 500 ft. ; 14,100 tons ; Speed, 23·3–24·1 knots, ; Completed, 1902–1903 ;
Armament, 2—9·2 in., 16—6 in., 12—12 pr., 7 small.

Above Plan and photograph of *Good Hope*.
Left The German cruiser *Gneisenau*.

Right The British warships *Invincible* and *Inflexible*.

The battle began when the two German armoured cruisers opened with a salvo at 12,000 yards (10,972 metres) which fell short. The next went over, then the third struck the *Good Hope* and the *Monmouth* and both ships burst into flame. The British ships stood gaunt and black against the reddish-yellow sunset, whilst their opponents were obscured by the evening mist and von Spee stood off to pound his opponents into blazing wrecks of splintered steel. His own ships received hardly a single hit in return.

The shells crashed into both ships, ripping away decks, blasting away funnels and turrets and spreading fire through their entire length. At 7.50 p.m., less than an hour after the first salvo had struck her, the *Good Hope*'s main magazine exploded. A 200-foot column of flame shot up into the air, brilliantly illuminating the grim scene below. Then as the debris splashed down again into the sea two of her guns, incredibly, fired a final defiant shot, then all was still. She was gone.

One of the *Monmouth*'s gun-turrets had exploded, but she was still afloat, although on fire and with only one gun in action. The *Nürnberg* came up and when her captain saw that the *Monmouth* refused to surrender, he opened fire on the helpless vessel at point blank range. The *Monmouth* slowly slid out of sight bows first, her stern, with the white ensign still flying, the last part to disappear beneath the waves.

The *Glasgow* seemed to have had a charmed life. Although more than 600 shells were fired at her, only one did much damage, and not in an area which could affect her speed. Realizing that she could do no more, she used her superior speed to outpace her pursuers and disappear into the gloom of the South American night.

The outcry in Britain at this defeat produced swift action. Admiral Sturdee was ordered to take command of the two new battle-cruisers, *Invincible* and *Inflexible* and to proceed at once to the Falkland Islands. They sailed with the minimum of delay and arrived there on December 7th. With them in harbour was the 11,000-ton heavy cruiser *Carnarvon* armed with 7·5-inch guns, the 10,000-ton cruisers *Kent* and *Cornwall,* the light cruiser *Bristol* and, with her wounds patched up, her sister ship *Glasgow,* all of which mounted 6-inch batteries. Also, in case von Spee had collected any auxiliary warships, the armed liners *Otranto* and *Macedonia* were included, giving Sturdee a three-to-one advantage in tonnage.

By some strange coincidence, the German fleet arrived off the entrance to Falklands harbour during the following morning, December 8th. Von Spee had intended to land a demolition party to destroy the wireless station when his look-out in the *Gneisenau*'s control tower reported the tripod masts of a battle-cruiser. As

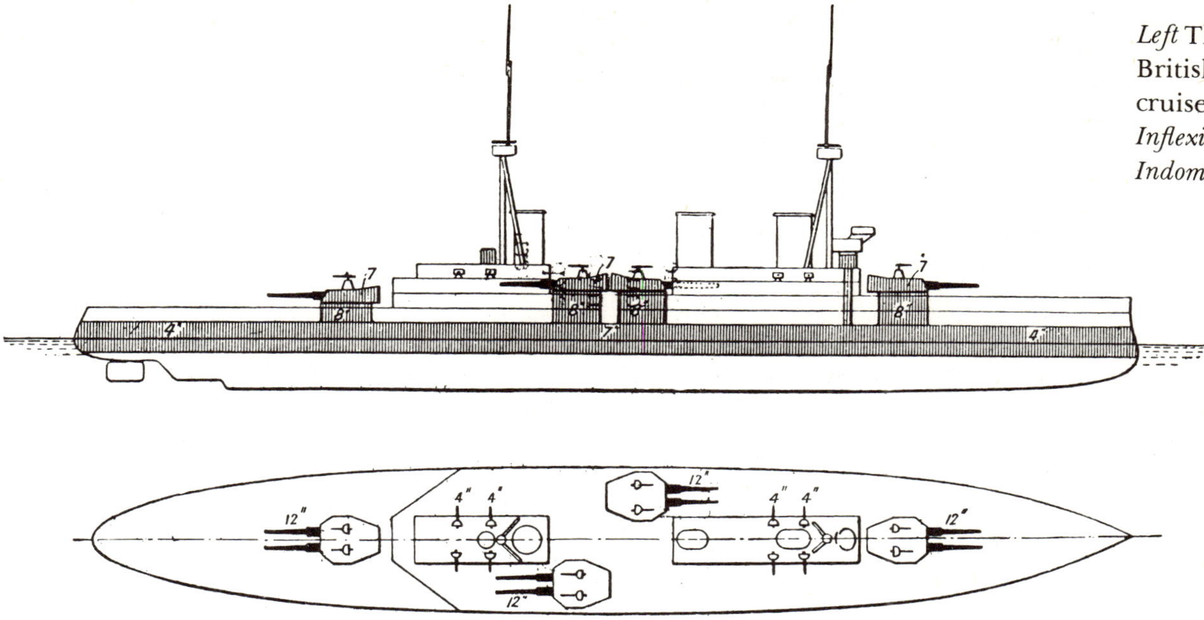

Left The design of the British armoured cruisers *Invincible, Inflexible* and *Indomitable.*

Invincible
Indomitable
Inflexible } Length, 530 ft. ; 17,250 tons ; Speed, 26 knots ; Completed, 1908–9 ; Armament, 8—12 in., 16—4 in.

* **Indefatigable** : Length, 555 ft. ; 18,750 tons ; Speed, 25 knots ; Completed, 1911 ; Armament, 8—12 in., 16—4 in.

* The centre turrets are more *en échelon* than in the three earlier ships.

the squadron drew nearer more tripod masts were seen. This could only mean one thing. At least two battleships, or battle-cruisers, had arrived at the Falkland Islands. There was consternation in the German ships and intense activity in the harbour as Sturdee ordered his ships to sail immediately.

Within a very short space of time the two large battle-cruisers and the *Carnarvon,* with the light cruisers close astern, raced out from their anchorage to chase the German squadron that was already making off to the south-east.

Had the chase begun later in the day, or had there been an overcast sky, von Spee might have escaped. But the British ships had all cleared the harbour before noon and were steaming in pursuit under a clear sky and upon a smooth sea. At 12.20 p.m., Sturdee decided to make the main attack with his three fast ships. With the sea piling up beneath their sharp bows, the *Inflexible, Invincible* and *Carnavon* moved smoothly ahead of the slower light cruisers and the range began to narrow rapidly.

At 12.55 the *Inflexible* opened fire at the *Leipzig,* the nearest enemy vessel. The *Invincible* also fired at the same ship but the range of 17,000 yards (15,545 metres) was too great and the shells exploded violently but harmlessly astern of the German cruiser. Soon a 12-inch shell threw up a tall column of white water so close to the *Leipzig* that she turned away in a south-westerly direction. Von Spee obviously wished to save his smaller ships if possible. He ordered the *Nürnberg* and *Dresden* to try and escape with her, but Sturdee immediately signalled his own three light cruisers to go in chase. His three bigger ships concentrated on the *Scharnhorst* and *Gneisenau,* although *Carnarvon* was some distance astern and frantically trying to catch up with the flagship.

The two German armoured cruisers suddenly turned to port but the British ships altered course, too, so that the four big warships were steaming on a parallel course, the range still decreasing. Although the weight of the British broadside was three times that of their opponents, it was the Germans who scored first; their crack gunners proved themselves better than the reservist gunners on board the *Invincible* and *Inflexible*.

The *Invincible* was the first to be hit. Sturdee then moved his ships so that he brought the German vessels within the range of his own 12-inch guns, but was outside the range of the German 8·2-inch. The *Scharnhorst* was hit by a salvo from the *Invincible* which started fires forward whilst the *Gneisenau* was being severely punished by the *Inflexible*. By 3.15 p.m. the former had one of her three funnels shot away but although the smoke hindered the British battle-cruisers, they used their speed to escape from it and continued to fire.

When the battle was at its height an amazing thing happened. A magnificent, full-rigged sailing ship appeared and began to sail majestically towards the battling ships. It was learned later that she was a French vessel which had sailed from Europe four months earlier. Without a radio her master had no idea that war had broken out whilst his ship had been at sea. She quickly went about and disappeared from sight. The thoughts of her master and crew as they saw the awe-inspiring sight of great warships, their guns blazing and tall columns of water rising near them from close misses, can only be imagined.

GUNNERY AND DAMAGE REPORTS ON BOTH BATTLES

	Rounds Fired	*Hits on ship*	*Killed*	*Power of Broadside*
CORONEL				
Good Hope	(?)	30–40	919	1,560 lbs
Monmouth	(?)	about 20	735	900 lbs
Scharnhorst	637	2	0	1,953 lbs
Gneisenau	442	4	0	1,935 lbs
FALKLAND ISLANDS				
Invincible	630	22	0	6,800 lbs
Inflexible	560	3	1	6,800 lbs
Scharnhorst	(?)	30–40	860	1,953 lbs
Gneisenau	about 1,000	30	580	1,953 lbs

At 4 p.m., with another funnel blasted away and with her hull blazing at several points, the *Scharnhorst* suddenly listed to port, rolled over, and with her battle ensign still flying slid slowly beneath the chill waters of the Atlantic ocean. By then the *Carnarvon* had joined her consorts and with a superiority of three-to-one the fate of the *Gneisenau* was settled. By 5.30 she had stopped, her outline lit redly by the flames that rose from the fires raging from bow to stern. At 6 p.m. she slowly disappeared from view for ever.

Meanwhile the three other smaller enemy warships were being chased by the *Glasgow, Cornwall* and *Kent*. The *Cornwall* and *Glasgow* concentrated on the *Leipzig* and soon after 7 p.m. she burst into flames, burned furiously for a while, then sank too. At the same time the *Kent* was busily chasing the *Nürnberg*, but was unable to close with her until two of the German ship's straining boilers burst. This slowed her down and she turned towards her opponent, prepared to fight to the bitter end. The *Kent* pumped shell after shell into her until at 6.30 the *Nürnberg* stopped firing and exploded into a giant bonfire.

The *Dresden* used her superior speed to get away and for months was a threat to merchant shipping. She was desperately hunted by ships of the Allied navies and for a time a grim game of hide-and-seek was played as British cruisers hunted their elusive prey. Not until March 1915 did *Dresden* reveal her position by a wireless signal and five days later she was sighted off Juan Fernandez Island. The *Kent,* in company with the *Glasgow* and the armed liner *Orama* raced in to finish her off.

But *Dresden*'s crew, short of ammunition and fuel and seeing that the end was inevitable, blew her up. She was the last remaining warship of von Spee's squadron.

Coronel had been avenged.

Chapter Four

The Somme Barrage

The Crash of Mighty Guns

During the first weeks of World War One it had been a war of movement. Armies advanced or retreated over an unspoiled countryside in the golden autumn of 1914. There were even old-fashioned cavalry charges with the French Dragoons, German Uhlans or British Lancers charging across open country in a style similar to that of their ancestors.

It was not long, however, before the war lost its momentum. The land on the Western Front was churned up by shell-fire and the movement of troops and wheeled vehicles, together with constant rain, turned it into thick, clinging mud. The men began to "dig in" and soon long lines of trenches were snaking away into the distance. On the Western Front alone they stretched for 475 miles. The fighting soldier was forced to exist under terrible conditions until the war ended in 1918.

Artillery at this time was classified generally as guns, howitzers and mortars. They may be further classified as: the *gun,* a long-barrelled long-range weapon. The path followed by its projectile (the trajectory) is relatively flat. The *howitzer* is a gun which fires shells upwards, on a more curved trajectory and is recognized by its short stubby barrel. The *mortar* has an even shorter barrel and a fairly short range because it has a trajectory shaped like a hairpin. They were usually fired from the trenches.

Below Long-range gun and *bottom* a howitzer.

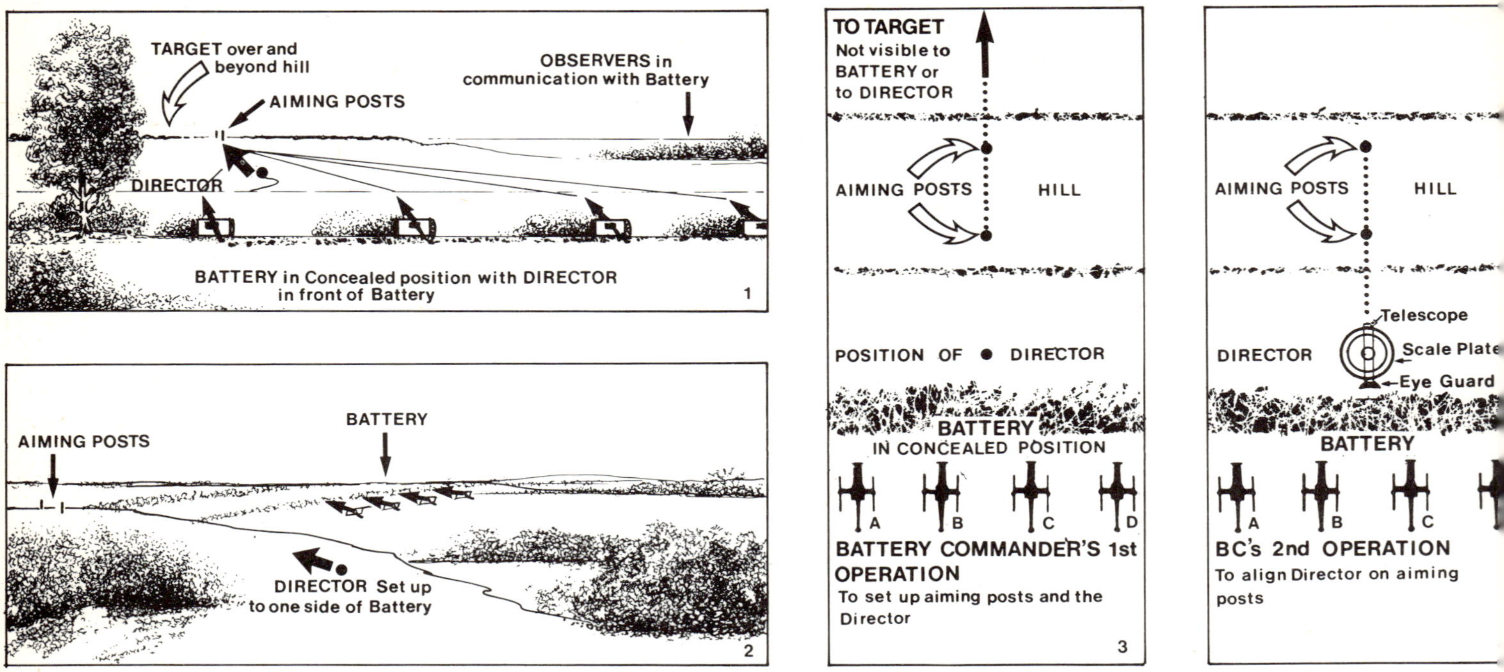

Above Gun-laying method in the Great War.

These were the sort of guns mostly used during the 1914–18 war; trench warfare did not call for much mobility in guns. Medium guns were used to pound trenches and close-to targets. The long-range guns and heavy howitzers were used to destroy enemy strong points and dislocate troop movements far behind the line.

Mortars
2-, 3- and 9·45-inch—up to 1,000 yards

Howitzers
6-inch 30 cwt howitzer—$2\frac{1}{2}$ miles
4·5-inch howitzer—3 miles
6-inch 26 cwt howitzer—$5\frac{1}{3}$ miles
15-inch howitzer—nearly 6 miles
8-inch (Mark VII) howitzer—$7\frac{1}{4}$ miles
9·2 inch (Mark II) howitzer—$7\frac{1}{2}$ miles
12-inch (Mark V) howitzer—$8\frac{1}{2}$ miles*

Guns
13-pounder—5 miles
18-pounder (Mark IV)—$5\frac{1}{2}$ miles
60-pounder (Mark I)—$7\frac{3}{4}$ miles
6-inch (Mark XIX)—nearly 10 miles
9·2-inch (Mark X)—over 15 miles*
12-inch (Mark IX)—over 19 miles*
14-inch (Mark III)—over 20 miles*
(*these were all on rail mountings).

Because many guns were fired over vast distances, some way had to be found for the gunner to know when he was "on target". A way of doing this had been worked out during the South African War. The system used was called "laying out the line": aiming posts were planted by the battery commander to direct the fire of one gun, whilst men from other guns in the battery would go forward with lengths of cord and plant their own posts which gave them a line along which to aim. This meant that the guns were all parallel. An observation post officer would observe the fall of shot from some raised ground and order the guns to be trained to the right or left as necessary.

This complicated and unwieldy system disappeared when the first director, fitted with a compass, was produced and used by the battery commander. The first directors were large and rather unwieldy, but they worked. There were times, of course, when the target was out of sight—behind a hill for example—and towards the end of the South African war balloons were used for observation. When the heavier-than-air machine began to be

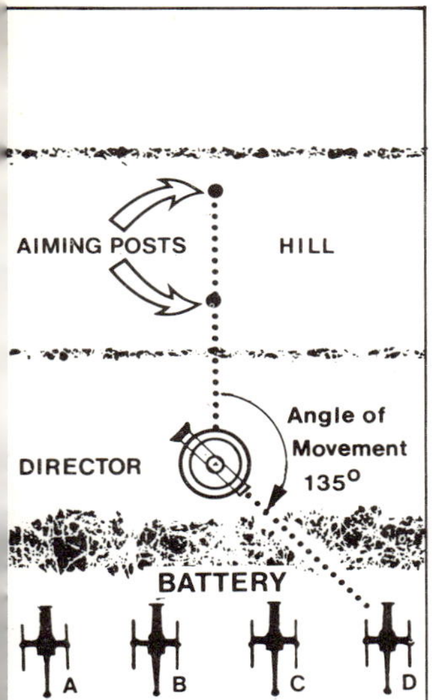

BC's 3rd OPERATION
To align Director on Guns (Diagram shows alignment on gun D)
5

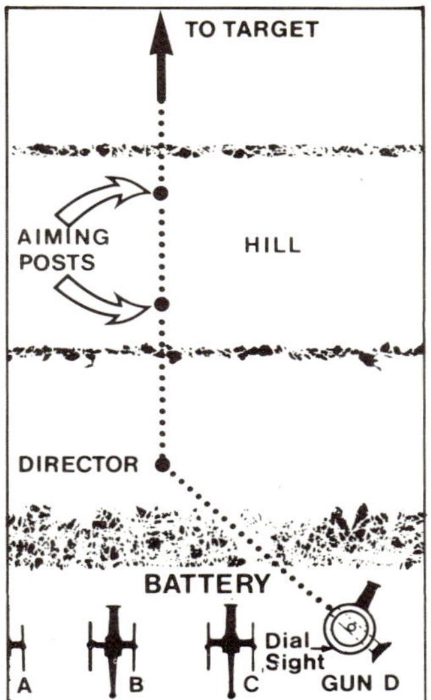

GUNNER'S 1st OPERATION
To align dial sight on Director (Diagram shows Dial Sight for gun D)
6

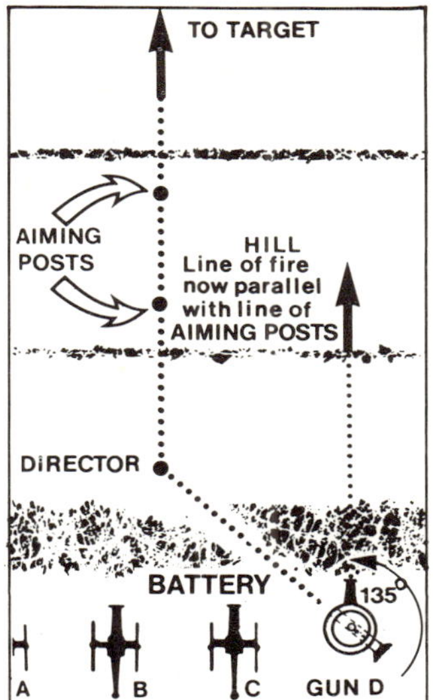

GUNNER'S 2nd OPERATION
To put gun (D) at position given by angle from BC's Director. See diagram 5.
7

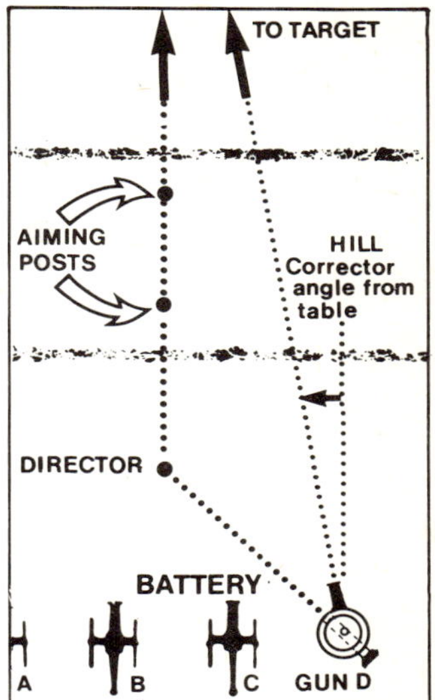

GUNNER'S 3rd OPERATION
To put gun on horizontal line of direction to target by means of corrector angle ascertained from table.
8

developed after 1909 it was obvious that it would make an excellent observation machine, although captive balloons were also used for "spotting" until the end of the war.

When the target could be observed from the ground, forward observation officers would shelter in the front line trenches, keeping in contact with their batteries in the rear by means of the field telephone.

The system of trench warfare varied from place to place, according to the terrain and the state of the ground. At first, straight and single trenches were dug but as the war progressed, they became irregular (traversed) in shape; this made enemy artillery-ranging more difficult. A network of communication trenches led up the "front line". These allowed reliefs, messengers and parties carrying ammunition and food to have some shelter for the last 400 yards or so. The British adopted a triple trench layout consisting of the front, the support and the reverse trench, each some 200–400 yards apart.

The front line trench had shell-proof dug-outs and was protected by hundreds of posts linked together with barbed wire. The three trenches were also joined together by communication trenches, dug in a zig-zag pattern. Along these ran the telephone cables to the artillery batteries and battalion headquarters in the rear.

The three major armies—British, French and German—deployed their artillery behind these trenches (with the exception of trench mortars). The field guns, which had maximum ranges of 6,000 yards, were set up in batteries some 2,000 yards behind the trench system. Medium howitzers were placed a little further back. Howitzers were particularly useful in trench warfare because their shells, falling in a steep incline, were very effective against the parapets of the trenches and deep dug-outs. The lighter field guns were used mainly for wirecutting and against an attack across the stretch of land between the two front line trenches known as "No Man's Land".

Below Typical trench design.

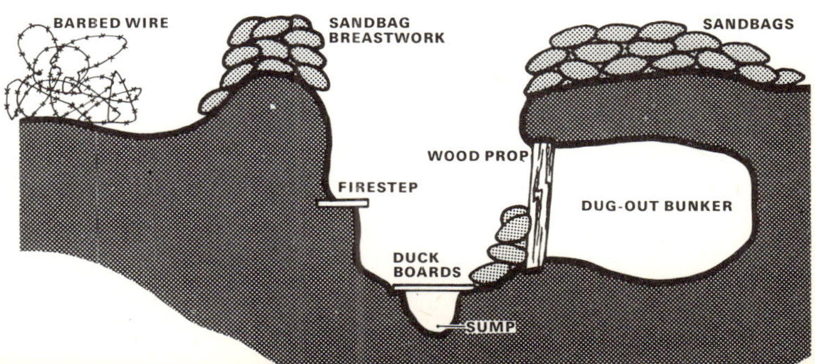

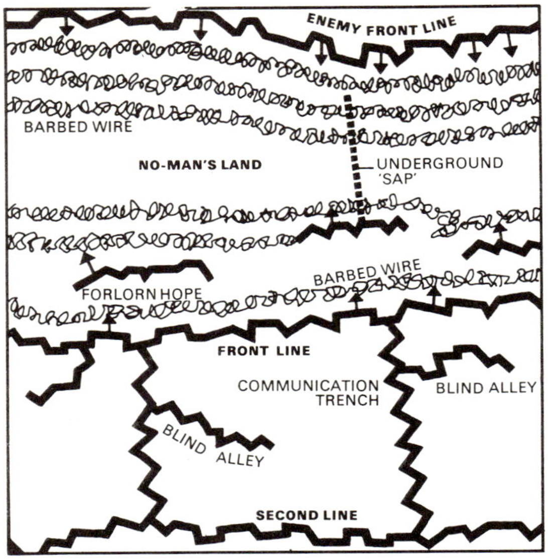

Right Typical trench lines.

The "heavies", the 60-pounders and 6-inch howitzers and above, were placed further back still. These were used mainly to shell troop concentrations behind the enemy trenches. The number of guns on the Western Front grew until in 1916, when the Battle of the Somme was fought, there was one piece of heavy artillery every twenty-nine yards "axle to axle" as Sir Douglas Haig had promised.

Below Mass production of howitzers.

By 1916 an immense number of soldiers were involved. Britain alone, thanks to conscription, had five million men in the field. In February of that year a great German attack was launched against the fortress of Verdun, the key to the entire French defence system. Confident of success the Germans moved up huge siege guns including their 42-cm "Big Berthas" and some naval guns which had a range of 29 miles. Although the French defences were inadequate for modern warfare, they also had some large guns with which to strike back, including 37-cm howitzers.

In order to relieve the pressure on Verdun, the Allied High Command drew up plans for an offensive on the River Somme. The British objective was to be the German-held town of Bapaume whilst the French were to make a direct assault on Peronne.

During the middle weeks of June 1916, a steady stream of guns was moved into position ready to begin the Battle of the Somme. The Allies began to assemble the greatest concentration of firepower ever known. Altogether there were 1,300 heavy guns, large quantities of medium artillery and, to feed their hungry charges, a stock of more than two million shells of all calibres was brought up and laid ready.

The plan—on paper—was extremely simple. This great mass of artillery was to start firing, and keep on firing, swamping the German trenches and rear defences with an avalanche of high explosive shells—for *seven days*. At the end of that time, the Allied generals decided, their infantry could climb out of their trenches to "go over the top" with confidence. Surely no one would be left alive to hinder their advance.

The main British armament consisted of the new 8-inch Mark VI howitzer, a monster weighing nearly $9\frac{1}{2}$ tons and having a range of 10,000 yards (9,000 metres or more than 5

Right from top to bottom: a British 60-pounder gun; an Austrian 28-cm. howitzer; and Sir Douglas Haig inspecting his troops.

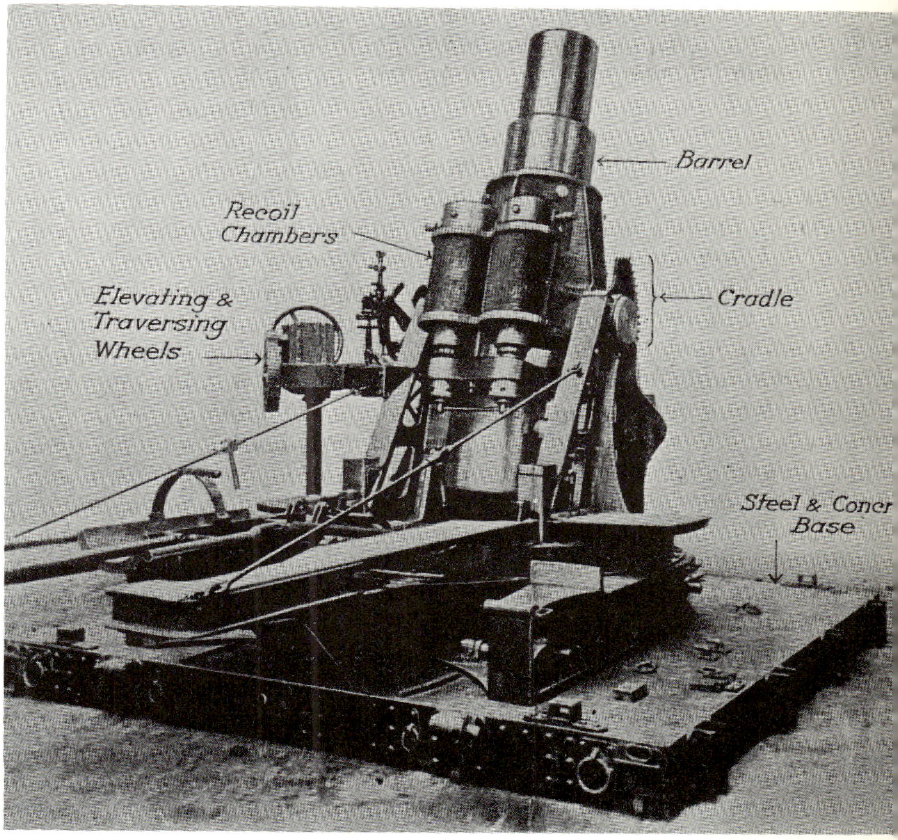

miles). Beside them stood older 9·2-inch and 8-inch howitzers and a new type of 6-inch howitzer which was half the weight of the latter but with a similar range. There were also a few 15-inch howitzers manned by the Royal Marine Artillery.

The battlefield at the outset of the battle was very different from the muddy coal-blackened soil at Loos and other places where battles had recently been fought. It was mainly agricultural, well-drained land which was perfect for trench digging. In front of the British trenches lay a series of chalk hills held by the Germans; they were strongly fortified and believed to be impregnable. Behind them a network of railways allowed reinforcements to be rushed to any threatened point whilst the German trenches to the front of the hills were deep and well constructed.

The British Commander-in-Chief, Sir Douglas Haig, was in no hurry. He was determined to make sure that everything was ready. As he said in one of his dispatches, "Scores of miles of deep communication trenches had to be dug, as well as trenches for telephone wires, assembly and assault trenches, and numerous gun emplacements and observation posts."

But at last even Haig was satisfied. At 7 a.m. on the morning of June 24th, 1916, the first heavy gun belched smoke and flame and the first great shell went spinning on its way towards the enemy lines. The British guns, parked in batteries of six to a battery (the French used batteries of four guns)

Top An enormous German mortar.
Left and below Setting the range of a mortar.
Below A German mortar and part of it crew.
Facing page (top to bottom): A French 25 mm. howitzer; a rail-mounted siege gun; and an inspection of a French mortar.

soon followed. The noise crescendoed to a wild frenzy as thousands of guns of all calibres began to join in the chorus of hate. The gigantic vibration was even felt in London. Observers in forward trenches watched the fall of the shells and signalled back to the battery commanders, continually adjusting the range.

To soften up the German trenches, a new system of warfare known as the rolling barrage (or concentration of fire) was used. The guns laid down this barrage upon a pre-arranged line, then after a while lifted to concentrate on one farther ahead. Also, when the infantry were to go forward, a similar barrage would be used, blasting at everything just before the soldiers and keeping pace with them as they advanced, hoping to destroy all opposition before they reached each objective.

That was the theory, anyway.

The British guns played havoc with the barbed wire defences strung before the German trenches and the fortified woods and villages beyond. British morale was very high. Everyone was confident that this was to be the great push, the last great endeavour that would bring the end of the war, and victory, in sight.

The great guns roared day and night, smashing into the soft earth and creating huge craters that slowly filled up with greenish-black and evil-smelling water. Unfortunately, no one had told the High Command that the drainage system of the area would collapse under the fantastically heavy barrage that was ripping the gentle, smoothly sloping country-side into mud and ruin.

At first the guns fired mainly at the barbed wire to give the infantry an unhindered advance. The targets changed during subsequent days so that the guns could pound the enemy trenches; the largest howitzers lobbed their shells over the chalk hills to pound the enemy support positions hidden from the gunners' sights. They were not firing entirely blind, however; spotting planes droned overhead protected by fighter planes of the new Royal Flying Corps. The smoke-grimed and sweating gunners would briefly glance up to watch the dog-fights being fought over their heads, then they would return to their immediate task of feeding the hungry guns with the shells that were spreading such destruction through the German lines.

During the early days of the bombardment the spotter planes sighted and reported back more than a hundred enemy batteries. These immediately became the target for the British howitzers.

The guns did not fire non-stop. There were periods when they would fall silent and then, thinking that an attack was imminent, the German infantry would emerge from their deep dug-outs to line the parapets of their trenches, ready to fight off the expected attack. Then a sudden burst of shell-fire would catch many of the defenders in an exposed position.

The crash and boom of the massed guns continued until men were half crazed with the sound, then at 7.30 on the morning of Sunday July 1st, the barrage lifted. Whistles blew, and confident that the greatest bombardment ever fired had proved completely successful the first waves of khaki-clad infantry went over the top. Each man was heavily laden with at least 66 lbs of equipment—ammunition, grenades, entrenching tools and heavy haversacks. Their progress across the churned up earth of No Man's Land was painfully slow.

For a while they moved forward without a shot being fired. Many believed that the enormous bombardment had wiped away all opposition. They soon had reason to change their

Above A farm seen from the air before bombardment.
Below The same farm from the air—after bombardment. The entire area has been pulped beyond recognition.

minds. The Germans, convinced that the long-awaited attack had begun at last, sounded the alarm signal along their battered but still intact trenches and machine gunners emerged from their deep dugouts, set up their guns . . . and waited. As one German officer later described the scene: "A few moments later, when the leading British line was within a hundred yards, the rattle of machine-gun and rifle broke out along the whole line of shell holes . . . Whole sections seemed to fall, and the rear formations, moving in close order, quickly scattered. The advance rapidly crumpled. All along the line men could be seen throwing up their arms and collapsing, never to move again. Badly wounded rolled about in their agony, and others, less severely wounded, crawled to the nearest hole for shelter."

It was not warfare, it was mass butchery. None of the targets was reached and by nightfall 60 per cent of the British officers and 40 per cent of their men had become casualties. Very soon the battle had become a stalemate, yet General Haig insisted on throwing fresh troops into the fighting, and the casualties mounted every day.

When the Battle of the Somme finally ended during November 1916, 418,654 British, 194,451 French and about 650,000 Germans had been killed or wounded. Despite the vast number of guns employed by the Allies, despite the fact that nearly two million shells were fired, the whole affair was a military disaster.

Almost a whole generation of young men had been killed, merely for a muddy wedge of shell-pocked ground that was, in effect, of no use to anyone.

Above A view of life (and death) under the big guns, by an artist of the Great War.

Chapter Five

Jutland

The Clash of Giants

Admiral Nelson was once approached by an inventor who wanted to interest him in a new gun sight that would allow accurate firing up to half a mile. Nelson replied that he always hoped to get so close to the enemy that his gunners could not miss!

Every one of his captains tried to manoeuvre his ship so that his opponent was virtually alongside. Each gunner would peer along the barrel of his gun, train it by means of a lever and then, having guessed the range, elevate the gun by means of a wedge beneath the breech. After the gun had been fired it had to be wheeled inboard by the crew hauling on the tackle, swabbed out with water to extinguish any sparks that might have remained; then the ball, cartridge and felt wad needed to keep the ball in place would be dropped down the muzzle and rammed home. The gun was then run out again ready for firing.

Left A dramatic picture showing the huge rise in naval firepower—Nelson's flagship *Victory* photographed by H.M.S. *Hercules*.

This was known as muzzle-loading. Curiously the warships of the twentieth century nearly went into action using the same out-dated method. In the 1860s MLs (muzzle-loaders) had a number of advantages over the BLs (breech-loaders). The breech-loading system wore badly and became unsafe. The shells often flew erratically if the breech piece was not properly closed. At this time, too, warships began to be armoured with iron plating (the "ironclads") and the BLs, with their weak charges, could not stand up to the heavy propellant charges necessary to send a shot smashing through such armour. Trials were held—MLs against BLs—and the muzzle-loaders won. This was because, unlike the earlier muzzle-loaders which had smooth bores, the new weapons were now RMLs (*rifled* muzzle-loaders). They had three grooves cut in the barrel and the shells had two rows of projecting soft metal studs which slid into these grooves. When the gun was fired these studs travelled up the grooves and the shell went spinning on its way.

For a time no further BLs were built and in 1871 the navy was equipping all her ships with RMLs of steel and iron.

H.M.S. *Devastation* on naval manoeuvres in 1887 in the Irish Channel.

Top A 100-pounder Armstrong naval gun of 1862.
Right A heavy cannon of Nelson's time.
Below Inside the gun turret of an ironclad of 1874.

The guns were built on Armstrong's principle of an inner barrel of steel surrounded by jackets of coiled wrought iron. This method of gun-making was a great improvement over the former crude casting. All RMLs were made this way with the addition of a solid forged breech plug.

The RMLs grew bigger and bigger. Some reached the incredible size of $17\frac{3}{4}$-inch calibre, firing a 2,000-lb projectile at something like 1,600 feet a second. Yet to achieve such results barrels became longer and soon difficulties arose in loading the guns, especially on board ships. Nevertheless, the standard muzzle-loaders of the time were 10-inch, 18-tonners which fired shells weighing over 400 lbs, and which could penetrate 10 inches of iron at 2,000 yards. Some were even larger, but even the 12-inch muzzle-loaders had a range of little more than two miles, the maximum range of H.M.S. *Victory* some seventy years earlier. After every shot they had to be trained inboard and their muzzles lowered into loading ports in the warship's deck.

They continued in use, however, until a terrible accident occurred in H.M.S. *Thunderer*. This ship was armed with two 12-inch RML guns. During a practice shoot her guns (mounted as a pair in one turret) were loaded with their 800-lb shells and trained outboard. The order to fire was given but one of the guns misfired, a fact strangely unnoticed by her crew. The guns were swung inboard for reloading and another shell was rammed into the gun—but it still had the first shell in the barrel owing to the misfire. The order to fire was given again and the double-shotted gun burst apart, killing almost every member of the turret crew together with a number of men who were standing on the battleship's deck.

The supporters of the breech-loading system at once claimed that such an accident would never have happened with a BL. Their argument carried the day. From 1881 on, breech-loading was adopted in all medium and heavy guns, military as well as naval, yet many RMLs continued in service until withdrawn for scrap.

By the beginning of the twentieth century every British warship was equipped with breech-loading guns. As maritime nations were quick to observe, they had rendered good service in the Japanese warships at Tsushima (1905).

In 1916, after two years of warfare, naval gunnery had already greatly improved. By then the standard guns in most battleships were 15-inch giants, mounted in pairs in an armoured turret. Each gun, with its mechanism, weighed more than 100 tons and was 54 feet 2 inches long. The turret mounting was protected by the barbette, a fixed cylinder of heavy armour which went from the turret down to the lowest armoured deck. The barbette was meant to protect the revolving turret structure which extended down to the ammunition magazines. The turret was revolved by means of power-driven gears and a toothed rack fitted within the barbette. The turret had to carry the enormous weight of the guns, the protective armour, and the ammunition hoist. The total weight of all this for a pair of 15-inch guns was 759 tons.

By 1916 the British Grand Fleet had been unable to force the German High Seas Fleet into general action. Although the Germans had made a number of "hit and run" sorties these were cowardly affairs. The big German warships slipped out into the North Sea, shelled peaceful seaside towns, then raced back to the safety of their harbours. The first sortie resulted in the bombardment of Yarmouth during November, 1914. A month later, Scarborough, Hartlepool and Whitby were ruthlessly shelled by battle-cruisers which were later intercepted by units of the British navy. Only a winter mist saved them from annihilation.

A further sortie in January, 1915, was also intercepted by British battleships. During the running fight that followed the German battle-cruiser *Seydlitz* was struck by a 13·5-inch shell which pierced her after-turret before exploding. The flash rushed down her ammunition hoist and only the prompt flooding of her magazine saved her from destruction. She managed to make port. There, experts put right an obvious fault, isolating the turret from the magazine and adding extra armour on the turret so that such a disaster would not occur again.

The British Admiralty, although aware of what had happened, did not take any steps to stop the same thing happening in their own warships. Consequently, when the two rival fleets were to meet later in the only major fleet action of the war, the British ships steamed into action as if with a large question mark hanging over them.

On the morning of June 2nd, 1916, people throughout Britain opened their morning papers to read: "On the afternoon of Wednesday May 31st, a naval engagement took place off the coast of Jutland." The article went on to list the British casualties which included the *Queen Mary, Indefatigable* and *Invincible,* together with the armoured cruisers *Defence* and *Black Prince.* Against this the communiqué could only announce vaguely that "the losses of the enemy are serious."

The whole nation held its breath until a fuller communiqué was issued. Meanwhile the German fleet, on returning to harbour, claimed a great and glorious victory over the British fleet, an announcement repeated by newspaper headlines all over the world.

When the full story of the battle finally appeared the British and their Allies were still left wondering. Was the Battle of Jutland a victory, a defeat, or a draw?

The battle had come about by a number of strange coincidences. Admiral Scheer, Commander-in-Chief of the German High Seas Fleet, wished to set a trap for the British battle-cruiser squadron commanded by Admiral Beatty which he knew to be operating in the North Sea. Scheer would send his own battle-cruisers under Admiral Hipper to make contact with their opponents. Then they would turn; Beatty's ships would be sure to follow and they would be led straight under the guns of his own powerful battleships.

It was a good plan, but it had one fault. Admiral Jellicoe, Scheer's opposite number in command of the British Grand Fleet, had intercepted some German wireless signals. Jellicoe decided to leave harbour with his own battleships ready to support Beatty if necessary.

On May 31st Hipper and five battle-cruisers, four light cruisers and 30 destroyers left the Schillig Roads to steer northwards. Some fifty miles astern came the great battleships of Scheer's fleet, 22 of them with six light cruisers and 31 destroyers. At almost the same time Beatty was sailing eastwards with his battle-cruisers *Lion, Princess Royal, Queen Mary* and *Tiger,* each mounting eight 13·5-inch guns, *New Zealand* and *Indefatigable* with eight 12-inch guns and, ploughing along some miles astern, the 5th Battle Squadron of four battleships, *Barham, Warspite, Malaya* and *Valiant,* each carrying eight 15-inch guns. Beatty's squadron intended to make a wide sweep and rendezvous with Jellicoe at 3 p.m. off the entrance to the Skagerrak. The combined British fleet would then total 28 battleships, nine battle-cruisers, eight armoured cruisers and 78 destroyers and, for the first time in

Right Map of the Battle of Jutland.

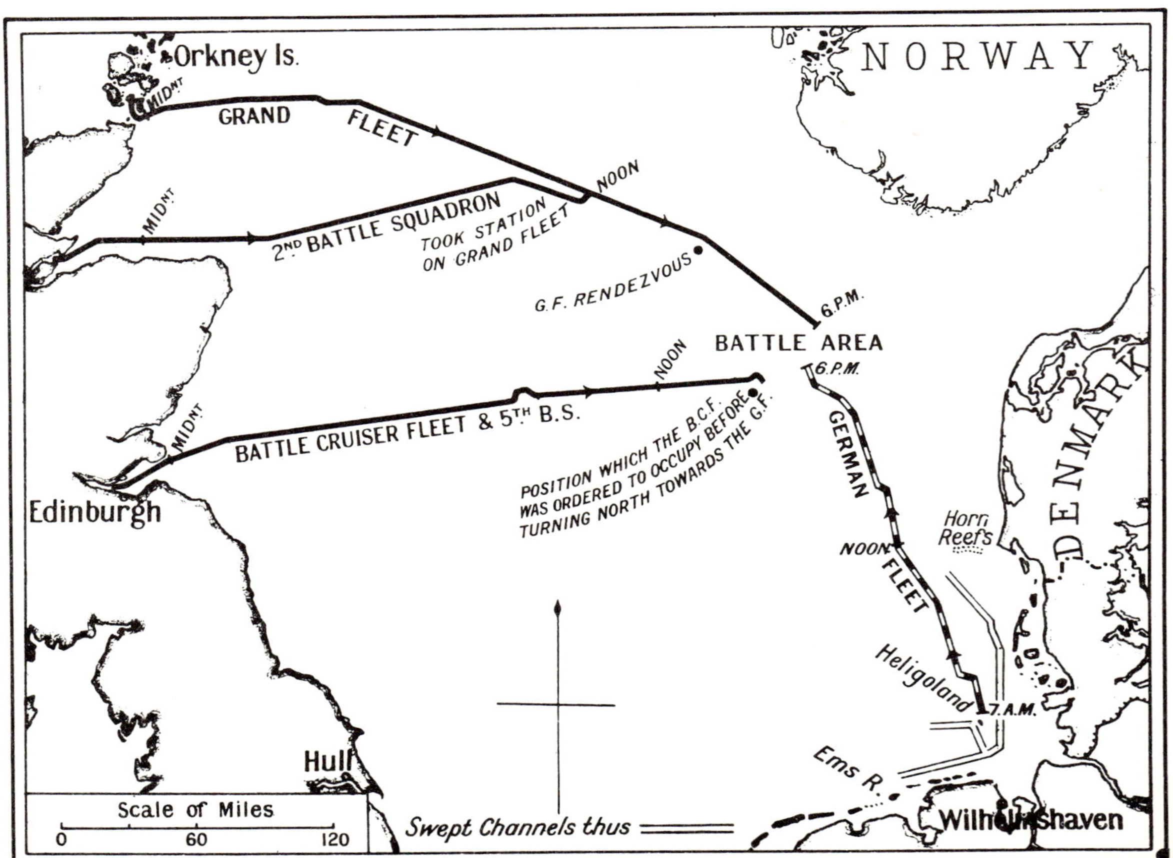

REPRODUCED FROM THE ORIGINAL OF R.D.CRAWFORD, BY PERMISSION OF "THE ENCYCLOPAEDIA BRITANNICA"

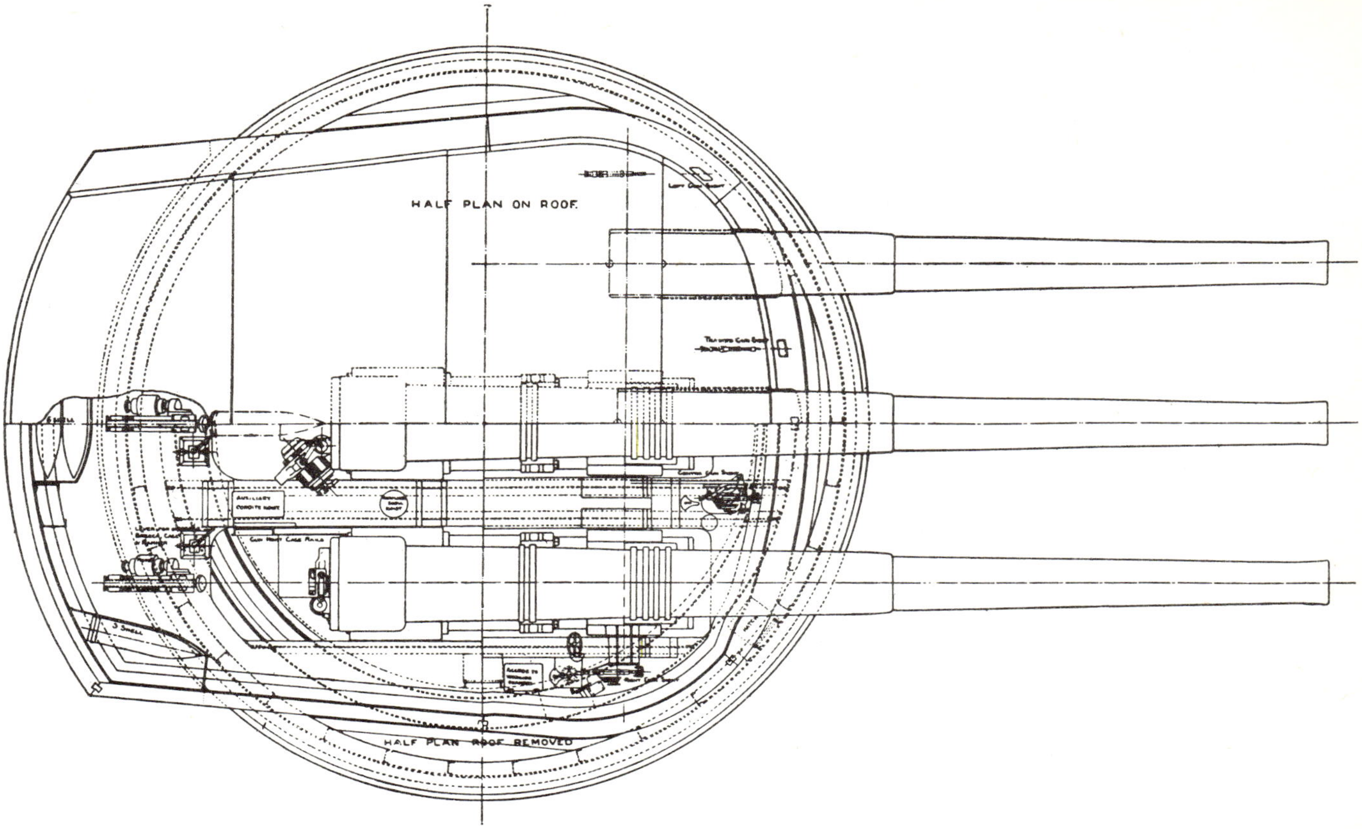

Plan and section of an Armstrong triple-gun mounting, of about 1913.

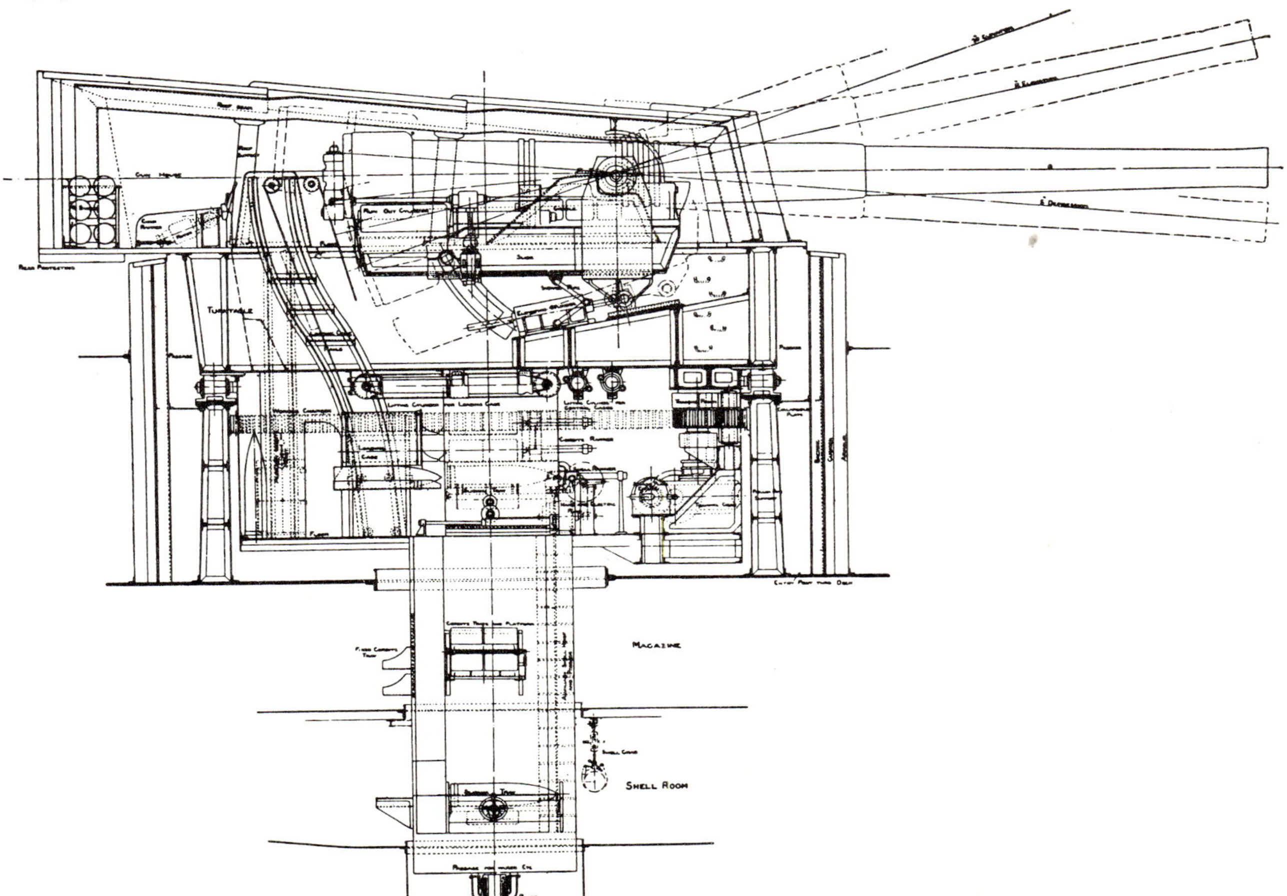

Above Sir John Jellicoe.
Right Admiral Hipper.

naval warfare, a seaplane carrier, the *Engadine*.

At 2.20 p.m. on that day, the light cruiser *Galatea,* scouting ahead of Beatty's squadron, sent back the nerve-tingling message, "Enemy in sight!" She had sighted one of Hipper's squadron off the Jutland Bank and a few minutes later both ships opened fire. At this time neither Beatty nor Hipper knew that the other's squadron was supported by a main battle fleet, although at that time they were still 100 miles apart.

The six battle-cruisers were soon engaged with Sheer's five. The air was shattered with the thunder of the guns, the whine of shells and the hiss of water as shot fell short. The sweating gun crews wheeled the ammunition from the magazines to the hoists. Others handled the shells as they arrived at the gun positions. Guns were loaded, the range and elevation were given and then each ship shuddered and heeled violently as its great guns belched flame and smoke at their opponents.

Both the *Lion* and *Tiger* were hit several times by superb German gunnery. Indeed, a shell exploded inside the *Lion*'s turret and there could well have been an even bigger explosion had not the turret commander, although dying, ordered the magazine to be flooded, thus saving the ship.

Once again, as at the battle of the Falkland Islands, an unsuspecting ship blundered into the battle area. This time it was the Danish steamer *Naesborg* which suddenly found herself between the furiously firing ships. She hastily turned, and steamed off as fast as her ageing pistons would take her.

The *Indefatigable* was having a hard slogging match with the *Von de Tann* when there came a sudden explosion. She listed heavily then, struck again by a further salvo, she turned completely over and disappeared beneath the waves. Not a single member of her 1,017 crew survived to tell what had happened. At 4.30 p.m. there came a further disaster. The 30,000-ton *Queen Mary,* Britain's newest and most powerful battle-cruiser, was fighting gallantly with two opponents when she too was struck by a series of salvos. A shell blew off the top of one of her turrets. The flash from the explosion shot down the ammunition hoist and the whole ship erupted in violent flame. This was followed by a vast upheaval of water, an enormous discharge of smoke and steam, and the ship went down almost immediately, split in two by the force of her exploding magazine. Only 18 out of a crew of 1,266 survived.

Left and below
Photograph and armament plan of the British armoured cruiser *Queen Mary*.

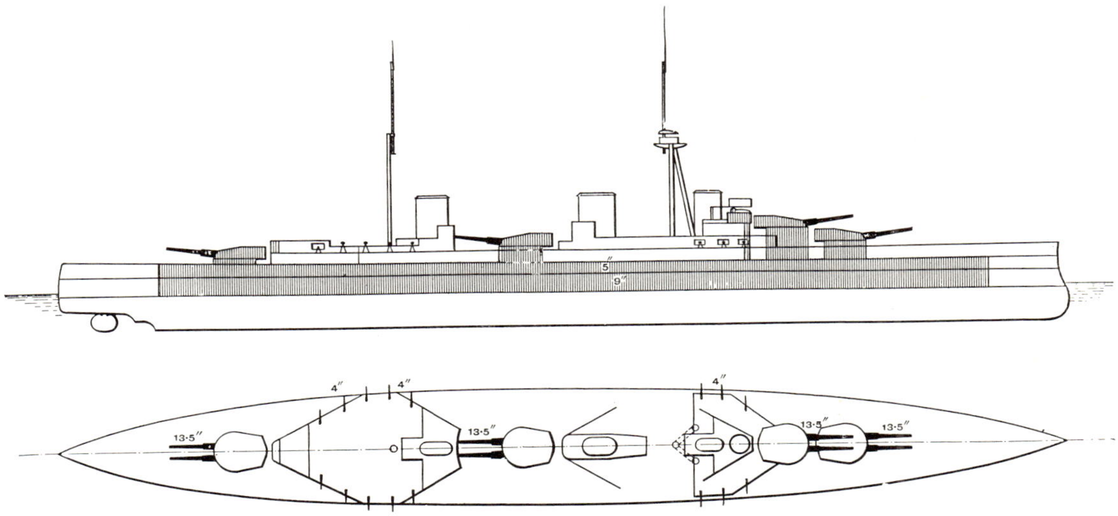

Length, 660 ft.; 26,350–27,000 tons; Speed, 28 knots; Completed, 1912, and Building; Armament, 8—13·5 in., 16—4 in.

Bottom The *Queen Mary*'s death throes at Jutland.

The unheeded lesson of the *Seydlitz* had resulted in both these losses.

Not long after this the battleships of the 5th Battle Squadron came into action but their guns were firing at very long range.

Shortly before 5 p.m. Admiral Beatty was startled to learn that the "heavies" of Scheer's High Seas Fleet were pounding towards him. He hastily radioed Jellicoe with the news then turned away, hoping to lead the enemy warships towards Britain's battle fleet steaming in from the east. Thanks to the superior speed of his battle-cruisers Beatty was able to draw away from Scheer's battleships but continued to fight a rearguard action with enemy cruisers.

At 6 p.m. the grim grey outlines of Admiral Jellicoe's mighty battleships began to appear on the horizon. This was as great a shock to Hipper as the earlier sighting of Scheer's fleet had been to Beatty. Neither had known that enemy battle fleets were in the area. For a time there was confusion as messages were only partially received or not received at all. Some ships, out of position, suddenly found themselves in great danger. One of these was the 14,000 ton cruiser *Defence* which found herself within two miles—almost point blank range—of Hipper's fleet. Within minutes the ship was hidden by towering spouts of water, then a number of well-aimed salvos turned her into a fiercely flaming wreck. Every man of the 907 crew lost his life.

Nearby, the *Invincible*, of Falkland Islands fame, was fighting a duel with the *Derfflinger* and thanks to superior markmanship was getting on top. Then disaster struck again. As an eye-witness reported: "I saw one of her picket boats go hundreds of feet up into the air, spinning like a leaf in an eddy of wind, followed by a huge lick of flame as high as her masts. A great belch of smoke, and then it was all over." She too had sunk.

By a strange coincidence, as at Tsushima, an eye-witness observer from another power was in the midst of the battle. He was Commodore von Schoultz of the Finnish navy, reporting for Tsar Nicholas II of Russia. Von Schoultz was on board the battleship *Hercules* with the main battle fleet and he later wrote about his experiences in his famous book *With the British Battle Fleet*.

Darkness was falling and the mist had thickened. It seemed that the fighting would die away for lack of targets. But suddenly the German battleships reappeared from the mist and the battle flared up once more. This time several German vessels were hit and when Scheer saw that some of the British ships were getting between him and his home port he gave the order to retire. Fighting still flared up when individual ships met each other in the darkness but by 4.30 the enemy had disappeared from the scene of the battle.

In terms of lost ships this was the final accounting:

BRITISH—*Queen Mary, Indefatigable, Invincible* (all battle-cruisers); *Warrior, Defence, Black Prince* (all armoured cruisers); *Tipperary* (flotilla leader) and seven destroyers. 14 ships sunk totalling 111,000 tons. Casualties 6,784.

GERMAN—*Pommern* (battleship), *Lützow* (battle-cruiser), *Elbing, Wiesbaden, Rostock, Frauenlob* (all light cruisers) and five torpedo boats. 11 ships sunk totalling 62,000 tons. Casualties 3,018.

Both sides claimed the victory, but the undoubted fact was that after Jutland, the German High Seas Fleet did not sail again until on November 21st, 1918, it was handed over to the British Fleet for internment at Scapa Flow.

From top to bottom: The *Seydlitz* in flames; the *Defence*; and the *Indefatigable* before Jutland.

Chapter Six

River Plate 1939

The End of an Ocean Marauder

When the Great War of 1914–18 finally ended, the nations that had been involved were left to mourn 10,000,000 dead and 20,000,000 wounded, an unimaginably frightening toll. Having destroyed Germany's formidable war machine, the Allies were determined that never again would the world be disturbed by her warlike efforts. Germany's army was reduced to 100,000 men, her navy to 15,000. She was not allowed to have an airforce. Every effort was made to see that she did not build up a secret military reserve.

Germany was forcibly disarmed, her guns, aircraft and warships were confiscated, and either scrapped or distributed among the victorious powers. Germany's days as a powerful nation were surely over. In 1920 the League of Nations was formed; the chief Allied powers (excluding the U.S.A.) were members. The Treaty of Versailles signed the same year banned Germany from building submarines and only allowed her to maintain a small token fleet to operate in the Baltic as a guard against attack by Russia. The Washington Treaty of 1921 prohibited Germans from building any new battleships larger than 10,000 tons.

The Allies also decided to limit their own armaments and for a time hardly any new building was undertaken. Germany remained quiet until the early 1930s when Adolf Hitler became steadily more powerful and urged his fellow countrymen to flout the Treaty. In 1934 when Hitler became Führer, he ordered the German army to be strengthened to 300,000 men and for the navy to be increased by the addition of new types of warships.

As if blind to all this, the Allies stuck to their own limited armament programme. France boasted the largest army in Europe, but she adopted a policy of defence, building a huge barrier of forts and gun emplacements stretching from Switzerland to the Ardennes Forest. Called the Maginot Line, it was armed with thousands of guns of all calibres. Its soldiers lived in air-conditioned compartments connected by underground railways. Yet when a British General inspected it early in 1940 he said: "It gives me but little feeling of security, and I consider that the French would have done better to invest the money in the shape of mobile defences such as more and better aircraft and more heavy armoured divisions than to sink all this money into the ground."

Yet Britain's "experts" were no better. Britain's army was allowed to run down; few new guns and tanks were built. Her navy, for many years the largest in the world, was also curtailed, its ships fast becoming outdated.

When the clash finally came with the outbreak of war in September 1939, it was found that many of Britain's warships did not match the quality of those of the enemy. In defiance of the Washington Treaty, Germany had been building ships far in excess of the 10,000-tons limit. By 1939 she had a number of powerful warships of an entirely new class. Officially known as armoured battleships—*Panzerschiffen*—they soon became known to the world as "pocket battleships". They were the *Admiral Scheer,* the *Lützow* (formerly *Deutschland*) and the *Admiral Graf Spee.*

Although technically "10,000 tons", they were nearer 13,000. Their 610-foot long hulls were packed with power. Their armament, too, was out of all proportion to their size; they carried six 11-inch guns mounted in two triple turrets, eight 5·9-inch guns, six 4-inch anti-aircraft guns and eight torpedo tubes. They also carried two aircraft which could be catapulted from their decks. Their armour was so thick that it was thought that no shells from British 8-inch cruisers could penetrate it, except at very close range.

These pocket battleships could outrun any battleship powerful enough to sink them, yet with their armour and firepower they could, in turn, overwhelm any ships fast enough to catch them. They posed a great threat to the Allies. In the whole of the combined British and French fleets there were only five vessels with similar power, the British ships being the *Hood, Renown* and *Repulse.* And these were already outdated.

British naval guns were built on the "wire wound" principle. Miles of steel wire were wound around a steel core to give additional strength to the gun

From top to bottom: The German warships *Admiral Scheer, Deutschland (Lützow),* and *Graf Spee.*

barrels. Although such guns were very safe from explosion, they were also very heavy. Guns of medium and lighter calibre were built up on an auto-frettage system. In this an outer jacket of steel was shrunk over the inner tube, making them much lighter than the wire-wound type.

The size of the main armament in the larger warships had increased to 15 inches by the end of World War One, but then the Washington Disarmament Treaty (1922) limited the size of naval guns to the 16-inch; yet the 15- and 14-inch guns were to become standard for battleships and battle-cruisers during World War Two. The 16-inch guns were huge, weighing more than 100 tons each without the breech mechanism, and firing a shell which itself weighed more than a ton. This huge shell was thrown on its way by a propelling charge of 640 lbs of cordite contained within a silk bag. Its range, like that of the 15-inch which fired a shell of 1,920 lbs, was more than 15 miles. The 12-inch gun of the *Dreadnought,* regarded as the last word, fired a shell of only 850 lbs, and the *entire* broadside of Nelson's flagship *Victory* in 1805 was only 1,160 lbs-less than one modern 15-inch shell!

The British Admiralty had at last learned the lesson of the *Seydlitz* and the reasons behind the heavy losses at Jutland years before. Three large battle-cruisers had been hit a few times each, but each of those hits was on a turret and the resultant flash caused the whole ship to blow up. By comparison the German battleships *König, Grosser Kürfurst* and the battle-cruiser *Derfflinger* were hit ten, eight and seventeen times respectively by heavy shells, yet still managed to struggle home.

These German ships had been fitted with anti-flash doors at points between the magazines, the hoists and the turrets. Also the silk-enclosed cordite cartridges were protected by metal cases until they were actually pushed into the guns' breech. The British cartridges, on the other hand, had been carried without any protection; they were exposed to the slightest spark. The British shortcomings were gradually put right in every warship.

Techniques in naval gunnery also improved. The system of director firing, particularly, had made great strides, and the element of guesswork entailed in watching the fall of shot had been reduced by mathematics and mechanization.

"Director controlled" guns were laid and fired by an officer working the director. This was a dummy sight placed high up to avoid smoke and to obtain a clear all-round view—from the foremast or from a lofty tower abaft the bridge. When the target was spotted, the director was trained onto it; all its movements were transmitted to pointers on dials showing elevation (height) and deflection (left or right training) situated in the gun turrets. The gunlayer in charge of the gun had only to "follow the pointer" keeping his own dials in step with the movements of the director-operated pointer, and so his gun on the unseen target.

As soon as the turrets had reported that their guns were loaded and ready for firing, the director officer made sure that his own sights were adjusted for range and deflection. Then he pressed a button that closed the firing circuit. Most British battleships in World War Two had four twin turrets and it was usual to begin firing with the right-hand gun of each turret in four-gun salvos. The left-hand guns followed at an interval dependent on the rate of firing.

The guns were aimed at the target by a complicated control system. The heart of this was the "transmitting station", a room well below the waterline which contained the fire-control table. This table was an ingenious calculating machine which produced a graph showing the speed and course of the target, and of the firing ship. Ship movements, the strength and direction of the wind and other factors which affect gunnery control were fed into the director, and this modified the graph accordingly. The director officer was kept supplied all the time with data from the transmitting station so that he could keep his sights always on the target.

The transmitting station also had an instrument which worked out the time of flight for the projectile from gun to target, allowing for the speed of the enemy vessel, and thus the exact moment when the ship's salvoes were due to fall. The "fall of shot" clock was invaluable when several ships were firing at the same target. Unless a ship was able to identify its own salvoes, effective gunnery control would have been impossible.

Radio ranging was also developed during the early days of World War Two and was to become a big factor in naval gunnery. A radio finder consisted of a complex electrical network which not only needed big electrical power supplies but also constant watching by trained technicians.

In the first few months of the war, however, such a sophisticated method was not in use, although both British and German ships had gunnery control equipment of the highest order. This was to be revealed when the Admiralty received news that the German raider, *Admiral Graf Spee* (which had slipped out to sea before war was declared), was operating in the South Atlantic, busily sending Allied merchant shipping to the bottom.

It was a matter of extreme urgency that the *Graf Spee* be dealt with. The nearest striking force of British ships was Force G, which consisted of the *Exeter, Achilles, Ajax* and *Cumberland*

Above The *Exeter*, and (*below*) the *Achilles*.

commanded by Commodore H. H. Harwood flying his flag in the *Ajax*. Harwood was ordered to "search out and destroy" the powerful ocean raider. Leaving the *Cumberland* to complete some repairs in the Falklands, he sailed at once with the other two ships. The stage was set for an interesting encounter between a single but very powerful battleship and three smaller but hard-hitting cruisers. Their respective details were:

SHIP	TONNAGE	GUNS	RANGE	SPEED
Exeter (The *Cumberland* was a sister ship)	8,309 tons	six 8-inch and four 4-inch AA guns	27,000 yards (15 miles approx)	31·25 knots
Ajax, Achilles (sister ships)	6,985 tons	eight 6-inch and four 4-inch AA guns	25,000 yards (14 miles approx)	31·25 knots
Graf Spee	10,000 plus	six 11-inch, eight 5·9-inch, six 4-inch AA guns	30,000 yards (17 miles approx)	27 knots

Dawn broke fine and clear on Wednesday December 13th, 1939, promising a clear day with good visibility. A little after 6 a.m. a cloud of smoke was observed on the horizon. The *Exeter* was sent to investigate and signalled back, "I think it is a pocket battleship," and a few moments later, "Enemy in sight bearing 322 degrees."

The position of the vessels at this time suited Commodore Harwood admirably. The enemy warship was coming down from the north-west with the *Exeter* steaming towards her on her starboard bow. The *Ajax* and *Achilles* were steering north-east, moving ahead of her to attack on her port bow.

Yet even as the ships closed, Harwood paced his bridge, deep in thought. He had many problems. He knew that the enemy's guns out-ranged his by more than two miles. He and his other two cruisers would be fired upon without being able to hit back. Only the *Exeter* had 8-inch guns whilst the *Graf Spee*'s armour was strong enough to withstand such calibre shells; he knew, too, that even the battleship's secondary armament of eight 5·9-inch guns were as powerful as the main armament on both the *Ajax* and *Achilles.*

Commodore Harwood was faced with two alternatives. He could either use the superior speed of his ships to close the range and attack at once, or use it to shadow the *Graf Spee,* keeping out of range until reinforcements arrived.

Harwood did not hesitate—he chose to attack.

At 6.18 a.m. the *Graf Spee* opened fire at very long range with her 11-inch guns, firing at all three cruisers at the same time, but at a combined speed of 60 m.p.h. the distance between them was shortening rapidly and the *Exeter* was soon able to fire her first salvo. As if this had been a signal, all the *Graf Spee*'s guns swung in her direction and the battleship's entire fire was aimed at her. Soon, one of the 11-inch shells burst close to her, damaging the communication system and starting fires. A minute later another scored a direct hit on B turret (the second from the bows), putting both of its 8-inch guns out of action. Metal splinters swept the bridge killing everyone except Captain Bell and two others. The gunnery of the *Graf Spee* was excellent—she mounted an early type of range-finding radar but did not rely on it during the battle. But Commander Graham of the *Exeter* tricked her into believing she was hitting his ship every time. Look-outs reported the flash of the *Graf Spee*'s guns and immediately depth charges were trundled over the side. Bursting at the same time as the enemy's shells, these made the German spotting officers think it unnecessary to alter the range.

All this time the *Exeter* was racing towards the other on a zigzag course. The trick was not discovered until the ships were less than four and a half miles away. By then *Exeter* had scored a number of hits on her opponent's upperworks which were not so powerfully armoured as her hull.

Because of the hit on her bridge, the *Exeter* was out of control for a short while until Captain Bell, conning her from the stern and using a boat compass and a chain of messengers, brought her under control once more. By then the *Ajax* and *Achilles,* coming up fast, had joined in with their 6-inch guns. The *Graf Spee* was forced to return their fire. The *Exeter* took advantage of this to alter course so that she might bring her port torpedo tubes to bear.

As *Exeter* turned, however, she received two more direct hits from 11-inch shells. One struck A turret (the foremost turret) the other entered her hull, doing severe damage and starting fresh fires. Y turret, the only one left in action, went on firing although the badly battered cruiser was now down by the head, more than 600 tons of water having flooded in through a hole in her bows. The fierce fires burning down below took a stronger hold. Many of the ship's hoses had been cut by shell splinters, but still *Exeter* struggled to keep up with the other two cruisers. She had been in action for 50 minutes and had been the *Graf Spee*'s main target for most of that time. 64 of her crew had been killed and many more injured, but she was still determined to stay in the action.

In the middle of the confusion caused by continual hits, with fires raging down below and water pouring in through her bows, there came a note of humour when it was found that the ship's pet, a canary, had laid an egg. Later when it hatched, the chick was raffled and a considerable sum was raised which was given to relatives of those killed in the battle. The chick was named "Graf Spee".

The fight went on, another 11-inch shell falling close to the *Achilles,* killing several ratings in the main gunnery control. But she raced on with the *Ajax,* the range closing all the time.

Then much to everyone's surprise, the battle became a chase. The powerful German battleship unaccountably presented her stern to her opponents, and under cover of smoke screens turned to the west. Once she made a sudden change of course and headed straight for the labouring *Exeter* as if to destroy her, but the approach of the other two cruisers made her turn and resume her original course, although she went on firing at her pursuers. It now became obvious that her captain, Hans Langsdorff, was running away. He ordered his ship to continue zigzagging and used smoke screens to try and protect her from the hail of 6-inch shells that were continually striking his ship, but both the *Ajax* and *Achilles* used these same screens, slipping in under cover of them and firing at close range.

Graf Spee's strength was still to be respected, for at 1.25 p.m. an 11-inch shell struck the after part of the *Ajax,* putting both X and Y (her two after turrets) out of action. Her speed had not been affected, however, and she turned to starboard, firing four torpedoes. Seeing them coming, the *Graf Spee* turned violently through 130 degrees and released yet another smoke screen. By now the two cruisers had closed the range to less than 8,000 yards and continued to pound their huge enemy with their guns until, finally satisfied that Langsdorff was about to seek shelter in the estuary of the River Plate, Harwood ordered his ships to cease firing and to drop out

Below Shot holes in the side of the German warship *Graf Spee.*

Above The *Graf Spee* begins to roll after being scuttled.

of range. He did not know what Langsdorff had in mind. With his battleship still a powerful fighting machine and with only two comparatively small cruisers to deal with, Harwood thought his opponent could have turned and faced them, fought them until nightfall, and then, if they were still afloat, escaped by disappearing with the darkness into the great wastes of the South Atlantic.

Harwood's two cruisers, with the *Exeter* struggling far astern, continued to shadow at a distance of 30,000 yards until, early in the morning of December 14th, the *Graf Spee* dropped anchor in Montevideo harbour. Ambulances rushed to the quayside and her 36 dead and some 60 wounded were taken off, her captain being amongst the latter, having been wounded in the arm.

Commodore Harwood sent the *Exeter* to the Falklands for urgent repairs and ordered the *Cumberland* to steam at full speed to join him. Meanwhile his two cruisers waited at the entrance to the River Plate like cats at a mousehole, except that the mouse was larger than both cats! Their vigil was later referred to as the "death watch".

According to International Law, a warship that is seaworthy may be forced to leave a neutral harbour and the Uruguayan authorities gave the warship until 8 p.m. on Sunday evening December 17th to leave or be interned.

Finally, just after 5 p.m. the *Graf Spee* weighed anchor and then most of her crew were transferred to the German supply ship, *Tacoma*. When the *Graf Spee* finally cleared the harbour she carried only a skeleton crew.

A few miles off the entrance to the River Plate the *Ajax, Achilles* and the newly arrived *Cumberland* waited. *Graf Spee* still had a considerable superiority over the three cruisers; all her guns were still in working order.

At 7 p.m. the battleship halted in the middle of the estuary and "experts" among the crowds were convinced that she was waiting until nightfall so that she could make a dash through the opposing screen of cruisers and so reach the open sea. But at 7.50 there came a terrific explosion. A great column of smoke and bright flame shot high in the air, then fires were seen bursting out along her entire length. As one onlooker wrote later, "At that moment the sun was just sinking below the

horizon, flooding the sky in which small grey clouds floated lazily, a brilliant blood red. It was a perfect Wagnerian setting for this amazing Hitlerian drama."

Amazed spectators on ship and shore saw life boats leaving the heeling, smoke-covered vessel. Captain Langsdorff was in one of them, the last man to leave the pride of the German Navy before the fuses detonated the tons of explosive that had been laid in her magazines.

It was an odd ending to the whole affair. Although the *Graf Spee* had come through the battle and still remained a perfect fighting unit, she had been scuttled by her own crew rather than risk another engagement at sea. Captain Langsdorff, unable to face the world after such a show of cowardice, locked himself in his hotel bedroom and shot himself.

What was the reason for such a defeat? How could one of the vaunted pocket battleships be out-manouevred and out-fought by three smaller vessels? It was because Commodore Harwood had followed the age-old maxim—"divide and conquer". As we have already read, the entire broadside of a warship is operated and fired by one man—the director officer, who can bring the greatest volume of fire to bear on the target. In the case of the *Graf Spee,* however, the same officer had an almost impossible task. By concentrating his fire on one ship, he ran the risk of undergoing the entire and undisturbed fire of the other two ships. When he divided his control and fired at two or even three ships, the work of his spotters was made much more difficult. Also, when he switched from one target to another, he lost accuracy: the range had to be worked out each time his guns ranged onto a new target.

The work of the director officers in the three opposing ships was far easier—they just had one target and just kept firing at it. This failure of his gunnery control was one reason why Captain Langsdorff broke off the battle. Later, misled by British propaganda which convinced him that massive reinforcements of battleships and aircraft carriers had arrived to join Harwood, Langsdorff decided to sink his ship rather than take his men into a battle which would cause the deaths of them all.

Above Torn apart by the explosion the *Graf Spee* sinks to the bottom of the estuary of the River Plate.

Above A massive rail-mounted 80-ton gun of the British army.
Right Two photographs of "Dora", the enormous rail-mounted siege gun used by the German army at Sebastopol.

Chapter Seven

Sebastopol 1941

An Epic Siege

As assault guns became larger, so did the problems of moving them from place to place. Originally the light and medium guns were hauled into position by teams of oxen, then by horses and finally by tractors. But some guns were too big for such mechanical aids. The dead weight of a gun is not the only reason why it is hard to move; this very weight can break up roads, collapse bridges and occasionally cause the gun to sink deep into the ground, making it well-nigh impossible to shift.

The modern answer to this is either to break the gun into sections and reassemble it at the point of fire, or to mount it onto a huge railway chassis.

The first method is used for guns such as the 8-inch. This, with its mount, is broken down into two sections so that its total weight of more than 30 tons may be more easily carried across country. Even so, its 34-foot-long barrel often causes difficulties when taken through a town with narrow streets or sharp turnings!

An even bigger gun, the 240-mm howitzer which fires a 360-lb projectile, is also dismantled into two pieces for cross-country transportation.

The largest guns are still moved by railway. The first time that a heavy gun was mounted onto a railroad wagon occurred during the American Civil War (1861–65) when a 13-inch mortar was used to bombard the city of Richmond, Virginia. The mortar consisted simply of a gun on a flatcar, firing through a sloping metal shield.

It was the 1914–18 war, however, which brought such guns into their own. Both France and Germany, with their dense networks of railways, used specially built railway guns; Britain utilized naval guns placed on specially built rail mountings. America, too, adopted the British method, stripping coastal defence guns from forts and old warships and manufacturing rail mountings to carry them. Some British rail guns used 12-inch guns taken from H.M.S. *Cornwallis* together with two 14-inch guns which had been made for Japan but were taken over by the British Army. The 14-inch guns were mounted on huge railway mountings and known respectively as "Boche-Buster" and "Scene-Shifter". Their projectile was a 1,400-lb shell fired more than 22 miles.

A great advantage of these railway guns was that much of the recoil could be taken up by letting the undercarriage run back along the rails until the normal recoil system came into play. The Germans cleverly took this a little further. They placed the gun just

outside a tunnel to fire, letting the recoil take it back into the tunnel so that the enemy spotters could not see where the shots were coming from.

The most famous gun of World War One was undoubtedly the Paris Gun. This was a specially built German gun of unique design mounted on a railway chassis and used to bombard the city of Paris 67 miles away. Curiously, the designers at Krupps used two calibres, a 15-inch barrel inside which was an 8·3-inch barrel. The latter projected 36 feet beyond the muzzle of the former. A further tube was screwed onto the 8·3-inch barrel to extend the gun a further 19 ft. 8 ins. so that the total length of the gun was 112 feet. It had a total weight of 138 tons.

The propellant was made especially for this unique gun, each round requiring some 350 lbs of power which was packed into a chamber nearly 16½ feet long.

The Paris Gun was brought up into position during March 1918, and opened fire on the morning of the 23rd. It caused terrific consternation amongst the people of Paris who thought, at first, that the damage was caused by high-flying aircraft. People could hardly believe that any gun was capable of firing nearly 70 miles. Yet this was so. The shells reached an astonishing altitude of 12 miles and travelled in an arc some 92 miles long in less than 3 minutes.

The gun went on firing during the summer of 1918, sending its last huge shell whizzing through the air on August 9th. By then the Allies were advancing steadily and threatened to over-run the gun. When they reached the site, however, there was no sign of it—it had been taken back to Germany where, it is believed, it was deliberately destroyed to avoid it falling into the hands of the victorious Allies.

Left "Dora" opening fire with a very steep trajectory.

Right (top to bottom): The legendary Paris Gun being moved into position and made ready for firing.

The 1939–45 war brought other similar giants into the field. Krupp's, the great German armament firm which had built the Paris Gun, now produced the K-12 Gun. This had a 105-foot long barrel and a range of 93 miles, outranging the Paris gun by 26 miles. Apart from a shell or two aimed at England it was never used and was fated to disappear from military history.

Another big gun of the time, and one which saw plenty of active service, was known by the surprising name of "Dora". It had an unbelievable calibre of 800 mm, or more than 30 inches, and weighed 1,350 tons. Dora was moved to its firing point in several pieces and put together on a specially laid railway track. This took more than six weeks. It fired two types of projectile, a 4-ton high-explosive shell with a range of 29 miles and a 7-ton concrete-piercing shell which had a range of 13 miles. It needed a crew of gunners, a maintenance unit and a guard totalling more than 1,400 men.

On June 22nd, 1941, when the war seemed to have reached a stalemate, an event occurred which was to shorten the war and help bring ultimate victory to the Allies. On that day Germany, Italy and Rumania declared war on Russia. Hitler's colossal military machine began to roll eastwards, over the Russian borders.

This eastern onslaught was called "Operation Barbarossa" and consisted of a massive attack by three powerful groups. One, the Northern Army Group, struck straight for

A Russian anti-aircraft battery during the German siege of Leningrad.

Leningrad. The Central Army Group headed for Moscow, the Southern Army Group for Kiev. With air squadrons flying overhead to bomb Russian airfields, the German invasion armies began to move forward on a vast front stretching from the Baltic to the Carpathians.

At first it was a *blitzkrieg* (lightning war) with tanks and light artillery smashing great holes in the Russian defences to let the following motorized infantry consolidate each gain. It was soon clear that the Russians had lost the battle of her frontiers as the German assault groups drove further and further into the vast Russian interior. Gradually, however, the defence began to harden, although by September, after only three months fighting, it was estimated that nearly two-thirds of Russia's fighting strength had been destroyed.

The war gradually became less mobile until, at last, the immense German armies ground to a halt with the approach of the cruel Russian winter. A number of important Russian cities still held out, and the Germans were forced to bring up their heavy siege artillery—the largest guns in the world. These began to bellow before such vital garrison cities as Leningrad (1941–43) and Stalingrad (1942–43). The siege of Leningrad, which lasted for sixteen months, was the most memorable of the whole war. That of Stalingrad served to drain the German army of its

manpower. Indeed, at the end of the siege of Stalingrad the entire Northern Group surrendered, 91,000 Germans were taken prisoner, whilst more than 120,000 had been killed in the fierce fighting. It was the turning point of the whole war.

Another important siege was that of the Crimean city of Sebastopol which led to its final abandonment, although the heroic stand by its defenders was to inspire the entire Soviet people.

The siege began in November, 1941, when German forces overran the Crimean peninsula and reached the outer defences of this great Russian fortress-port. Two attempts to carry Sebastopol by storm cost the Germans more than 60,000 dead; no further attacks were launched until the following June although air raids on the city went on day after day. This interval allowed the Germans to bring up their heavy siege artillery, guns which were the largest ever seen on any battlefield. There were 120 batteries which included 56 batteries of guns ranging from 190 mm to 420 mm, the latter being 20 mm larger than the largest gun ever used in the First World War, the French 400-mm St. Chamond.

In addition the Germans had a battery of super-heavy 615-mm mortars and their rail-mounted "Dora".

The German XI Army at the siege of Sebastopol was commanded by General Manstein, a most experienced soldier. Manstein was able to concen-

"Dora" with its crew, on its way to action at Sebastopol.

trate more than 200,000 men, nearly 670 big guns, 450 huge mortars and 720 tanks. Against this the Sebastopol garrison could only muster a little more than half the number of men, 700 guns (although these included some large naval guns used as coastal defence), some mortars and about three dozen tanks. Their aircraft were only 55 compared to the 600 of the *Luftwaffe*.

But the defenders had a fantastically high morale. Lt.-General Dietmar, a German military commentator, stated that the task to reduce Sebastopol "was certainly no easy one". Any German soldier at the siege would have put this in stronger language!

The Germans needed the port of Sebastopol to ease the supply problems of a major offensive in the south. They were determined to capture it at all costs. The Russians were equally determined that they wouldn't.

The invaders' superiority in tanks mattered very little for the countryside was quite unsuitable for their use. The approaches to Sebastopol consisted of a tangle of deep ravines and steep-sided gorges which not only made tanks useless but also prevented direct shelling. This type of ground also gave excellent cover for the defenders. They fought from positions deep in the rock, whilst large natural caves gave roomy shelter for both men and materials. Ammunition and food had to be carefully guarded and used, for everything had to be brought in by sea, which the Germans controlled. German motor-torpedo boats effectively cut off the main lines of communication. Towards the end of the siege, Russian submarines were even used to bring in desperately needed supplies to the Sebastopol garrison.

June 2nd saw the beginning of a tremendous artillery bombardment. Manstein's heavies and super-heavies ripped into the forts that encircled the city, blasting chunks of concrete and steel high into the air, dislodging guns and killing not only the men of the garrison but many of the civilians who had chosen to remain. This incessant bombardment went on until the morning of June 7th when the infantry were ordered into the attack, advancing under air cover provided by the *Luftwaffe* (nearly 1,000 sorties were carried out on the first day). Meanwhile the guns continued to hurl their enormous shells at the city.

Despite this combined assault the defenders stood firm, Russian artillerymen and tank crews often taking part in hand-to-hand fighting.

After a time the air raids on the city had become so frequent that warnings were no longer given, yet the 60,000 civilians somehow managed to carry on their lives despite the shelling and the bombing. Most of them lived underground where they had managed to create another "city"—a strange world where canteens, factories, newspaper offices and even schools carried on more or less as usual. Above them, the real city of Sebastopol was being smashed to rubble.

With ammunition running short the end was inevitable. Forts Stalin, Molotov, Lenin and Maxim Gorki were finally carried by assault. A German eye-witness said: "Fort Maxim Gorki went on fighting even after it had been captured. It sounds unbelievable, but it is true. Although the upper storeys of this fort are in our hands and the battle line has moved some 1,400 yards forward, Soviet soldiers are beneath us, deep underground in the lower storeys of the fort, and continue to put up resistance. We have sent negotiators to explain to them that further resistance is useless, but they refuse to listen; they will not come out. While we succeeded in forcing several hundred Soviet soldiers out of the upper storeys, blasting them out with handgrenades and explosives hurled through the apertures, we cannot reach these last men.

"Never before have we seen such armour-plate, such concrete walls and, above all, such a kind of concrete which is quite new to us. Fort Maxim Gorki had 13-inch guns which continued to blast away even after our shock units and storming guns had been brought up to close range. They went on firing at a range of less than 800 yards, and even 500 yards. We have still to deal with other forts. One, a very large artillery bastion, has been bypassed by our troops, but continues to hold out in our rear. We still have to silence heavy Soviet coastal batteries, the huge guns of which have been turned inland and shell us incessantly."

Fort Malakhov was stormed on June 30th and a rapid advance also took the harbour and town of Balaclava, a name well known to every Briton because of the famous Charge of the Light Brigade many years before. The next day the Germans were inside Sebastopol itself. By July 3rd the city was in German hands after a fiercely fought siege of 250 days. By then the civilian population had been evacuated by warships of the Russian Black Sea fleet. At last, when the end was imminent, a Dunkirk-like evacuation took place with minesweepers, patrol boats and submarines taking away many of the defenders under constant air attack. Others managed to break out to reach the mountains to carry on the war as partisans.

Never before had a fortress suffered such a prolonged assault or been blasted by such a colossal weight of armament. Seldom in history has the gallant fight put up by the besieged served such a strategic purpose, for it disarranged and weakened the German High Command's plans for a great offensive. Even more important, it helped to buy time desperately needed to allow Russia to mass her strength for the years that were to follow.

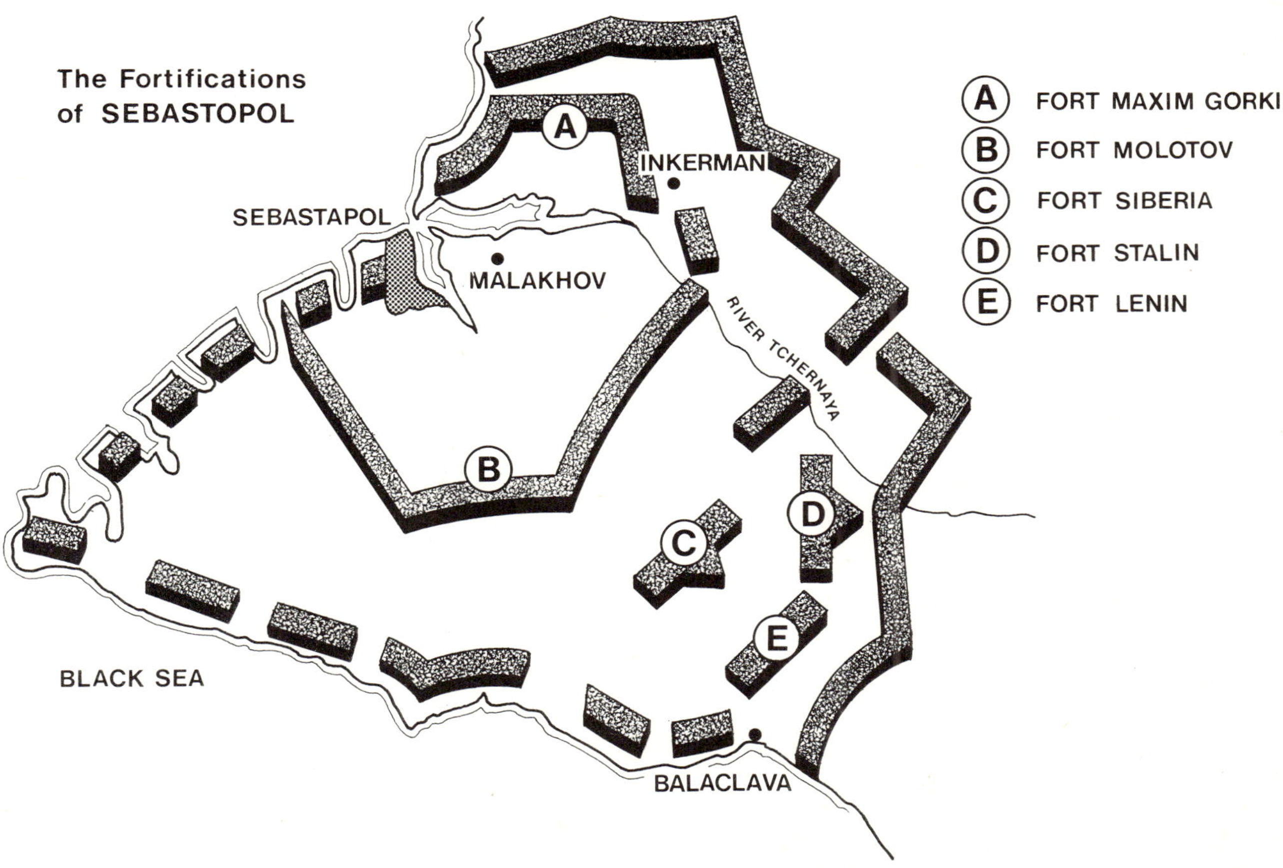

Chapter Eight

Leyte 1944

The Battle of Surigao Strait

The Washington Treaty which limited the size of warships to 35,000 tons ended in 1936. An agreement made the year before stated that the calibre of naval guns should not exceed 14 inches. Nevertheless Japan, since Tsushima (1905) one of the world's leading maritime powers, had not signed this agreement and so the U.S.A. decided to wait until the expiration of the Washington Treaty. Then, if Japan had still not signed, she would arm her new ships with guns larger than 14 inches.

Honouring the agreement, Britain built five new battleships, the *King George V, Prince of Wales, Anson, Howe* and *Duke of York.* All carried a new type of 14-inch gun which was superior to the old 15-inch both for its rate of fire (two rounds a minute) and for its power of penetration (13 inches of armour at a range of 15,000 yards). It had a range of 36,000 yards. The armament of these ships was ten 14-inch guns, sixteen 5·25-inch dual purpose and many anti-aircraft guns.

At the same time America was building four new battleships—the *Iowa, New Jersey, Missouri* and *Wisconsin.* They were the largest of all Allied battleships, being armed with nine 16-inch guns, twenty 5-inch guns and a multitude of 40-mm and 20-mm anti-aircraft guns.

These warships had a displacement of 57,600 tons but were exceeded by two enormous Japanese ships, the *Yamato* and *Musashi.* These had a displacement of 64,170 tons and mounted nine 18-inch guns in triple turrets, two forward and one aft. They also had eight 8-inch guns in twin turrets and twenty-four 25-mm guns in triple shields. They were massively built, their armoured sides being invulnerable to 18-inch shells at 20,000 yards. Their decks could withstand a 2,000-lb bomb dropped from a height of 10,000 feet. Despite their size and weight they were capable of a speed of more than 27 knots. This was the ultimate in battleship design.

Britain's final and greatest battleship was H.M.S. *Vanguard* which was not completed until 1946, a year after the war had ended. Her 15-inch guns had originally been part of the armament of H.M.S. *Courageous* and H.M.S. *Glorious* of World War 1; she had 14-inch thick armour and four separate engine rooms. Her armament was eight 15-inch guns, sixteen 5·25-inch and forty 40-mm anti-aircraft guns, yet even before she was launched it had become obvious that the day of the big gun battleship was over.

Air power had decided that.

The first aircraft carrier was H.M.S. *Ark Royal*; she was a merchantman converted for the purpose and used her holds for stowing sea-planes which

Facing page: top the United States carrier *Iowa* launched in 1943, and *below* the *Missouri* firing a salvo from its forward 16-inch guns.

were hoisted in and out by cranes. These float-fitted sea-planes could only take off and land on calm water which meant that their use was very limited.

At the outbreak of World War One in 1914, the old cruiser *Hermes* and six fast cross-Channel steamers were converted to aircraft carriers with four seaplanes each, hangars and cranes. They were armed with four 12-pounder guns and one of them, H.M.S. *Engadine,* was present at the Battle of Jutland, the first aircraft carrier to take part in a naval battle. Actually the part she played was a very minor one for although she located the German High Seas Fleet, she was unable to report her findings owing to a radio fault.

As the 1914–18 war progressed and the value of air scouting became more apparent, larger carriers were needed. One was the Cunard ocean liner *Campania.* She had a forward flying deck from which seaplanes were launched with wheeled trucks beneath their floats, the trucks being detached once the machine was airborne. It was soon realized that wheeled aircraft were far superior to seaplanes but there was the problem of landing on a carrier's deck although an American, Eugene Ely, had shown the way. Ely flew off an aircraft from a platform built on the bow of a cruiser as early as 1910 and two months later he landed a plane on a platform built on the stern of the U.S.S. *Pennsylvania.*

By the end of World War One in 1918 Britain had quite a big carrier fleet including H.M.S. *Hermes* (10,850 tons) the first ship to be laid down as a carrier rather than be converted from a cruiser or merchantman. *Hermes* was so unusual in appearance that she was nicknamed "Noah's Ark" and "The Covered Wagon".

The British aircraft carrier *Hermes*.

The effect of ship-borne planes in World War One was of little consequence and it was not until World War Two that they really showed what they could do. In 1939 when the war began the strength of the opposing fleets was as listed below:

In addition there were, of course, minelayers, escort vessels, tenders, gunboats, motor torpedo boats and others.

The effect of the striking power of aircraft in World War Two was vividly shown by the Royal Navy's victory at Taranto in November 1940, the partial destruction of the U.S. Pacific Fleet at Pearl Harbour on 6th December, 1941, and the sinking of H.M.S. *Repulse* and *Prince of Wales* by Japanese torpedo-carrying aircraft four days later. The construction of aircraft

Country	*Battleships*	*Battle-cruisers*	*Aircraft & seaplane carriers*	*Cruisers*	*Destroyers*	*Submarines*
Britain	12	3	9	64	179	58
France	8	2	1	50	28	72
U.S.A.	15	–	5	36	181	99
Germany	–	5	–	8	22	57
Italy	4	–	–	22	59	109
Japan	10	–	6	38	113	93

The Japanese air attack on the American Pacific fleet anchored in Pearl Harbour.

carriers was speeded up as the war went on. The Royal Navy laid down the *Illustrious* class of 23,000-ton carriers with a complement of 16,000 men. These carriers had seven decks which, from top to bottom were (1) flight deck, (2) upper gallery deck, (3) upper hangar deck, (4) lower gallery deck, (5) lower hangar deck, (6) upper deck and (7) main deck.

To operate a large number of aircraft they had to carry an immense amount of aviation spirit. And so any carrier was a floating fire hazard. The main supply of petrol was stored in cylinders in the hold at either end of the ship, with pipes leading to fuelling points in the hangers and on the flight deck. These cylinders were so arranged that they could be dumped overboard and sunk in an emergency. Fire pumps, sprinklers and chemical foam showers were also fitted throughout the ship ready to be brought into action in seconds.

Because of their open construction and medium armour, aircraft carriers were vulnerable to gunfire, mines, torpedo and aircraft attacks and so for most of the war the battleship still ruled the waves.

Gunnery techniques were constantly improving, although the method of observing the fall of shot (to correct the range and elevation before the next firing) meant human errors. Most probably the sky was overcast, with a big sea running and the ship rolling heavily, digging in deeply and sending showers of spray high over her bows. The enemy ships would probably look no more than smudges on the horizon. The speed of one's own ship and that of the enemy had to be calculated, the range, the wind force and the roll of one's ship had to be taken into account including the time it would take the shell to reach the target.

Britain's introduction of radar (Radio Detection And Ranging) was to change this hit-and-miss method. This was just as well, for it had been found that visual range-finding could be more than half a mile out!

By the middle years of the war, most Allied warships were fitted with radar gunnery control equipment. A few were equipped with an advanced radar gunnery control system known as centrimetric radar. This not only registered the fall of shot on its screen but actually measured the distance of each splash from the target. Thus an immediate and extremely accurate spotting correction could be made, rather than having to rely on the former "short, over and straddle". Radar had a further great advantage—it could be used at night where, previously, the fall of shot could not be seen and spotters had to rely on the flashes of the enemy's guns.

It was with such improved equipment that the last great naval battle of World War Two was to be fought between the American and Japanese navies in the Philippine Sea.

On December 7th, 1941, an attack by Japanese aircraft on the American

overleaf: The Japanese battleship *Fuso*, one of Admiral Nishimura's casualties at Leyte Gulf.

A British radar operator.

Pacific Fleet lying at Pearl Harbour marked a new phase in the war. Three days later the Philippines were invaded by the Japanese. A series of defeats compelled U.S. forces to evacuate the area although General Douglas MacArthur made a famous promise that he and his men would return.

This promise was kept when on October 20th, 1944, American forces swept ashore at Leyte Gulf. The Americans knew that the Japanese would do everything they could to stop this landing. Two powerful fleets, the U.S. 7th and 3rd, stood nearby to tackle any opposition. The Japanese, who also had four striking forces in the area, had been forced to disperse, not knowing where the Americans would invade. When Japanese land forces on Leyte radioed for help, separate actions were to be fought in the Sibuyan Sea (October 24th), in the Surigao Strait (24th-25th), off Samar (25th) and off Cape Engano (25th-26th).

Of the four phases of the Battle of Leyte Gulf, as it was later called, that of the Surigao Strait was the most interesting. It was not only the last naval big-gun battle of the war, it was also the last in history. For the most part the contending forces clashed as separate units of Task Groups or Striking Forces, but an indication of the total strength of both sides was:

	Battleships	*Aircraft Carriers*	*Heavy Cruisers*	*Light Cruisers*	*Escort Carriers*	*Destroyers*
U.S. 3rd Fleet	11	17	7	14	–	75
U.S. 5th Fleet	7	3	4	18	18	48
Japanese Fleet	9	4	13	5	–	31

Shortly before noon on October 24th, two Japanese Striking Forces were sighted by U.S. carriers' planes. The van commanded by Vice-Admiral Nishimura consisted of two battleships, the *Yamashiro* and *Fuso*, the heavy cruiser *Mogami* and four destroyers. Some miles astern came Vice-Admiral Shima's force of two heavy cruisers, the *Niachi* and *Ashigara*, the *Abukuma*, a light cruiser and a further four destroyers. They were all due to rendezvous with the 1st Striking Force A under Vice-Admiral Kurita, which had the two largest battleships in the world. These were the *Musashi* and *Yamato*, each armed with the world's largest naval guns, nine 460-mm (18·1-inch) in three triple turrets. Each of the eighteen guns weighed 162 tons and fired a projectile of 3,200 lbs.

This force, however, had been intercepted by Task Forces of the 3rd Fleet in the Sibuyan Sea and forced to withdraw. The "unsinkable" *Musashi* was sunk, but it took 19 torpedoes and 17 bombs to do it. And so this fleet was destined not to make the rendezvous.

It was decided by the U.S. Command that the two Japanese squadrons were on their way to attack the landing craft and transports at Leyte Gulf. So the whole of Task Groups 77–2 and 77–3 of the 7th Fleet were alerted to receive them. These were under the command of Rear-Admiral Oldendorf who was able to mount a massive force of battleships, heavy cruisers and light cruisers and four divisions of destroyers.

He also had 39 torpedo boats (known as PTs) and he sent these out on a night patrol of the entire strait and its approaches by way of the Mindanao Sea. The first boat picked up the leading Japanese force on her radar and with the other two boats in her patrol raced in to the attack. They were driven off by heavy gunfire but were able to radio their sighting to Olden-

dorf. The same thing happened right across the Mindanao Strait. Oldendorf was alerted as each group of boats tangled in turn with the advancing warships before retreating.

After Nishimura had steamed past this active PT screen, he came under attack from the first division of the U.S. destroyers. By now it was quite dark. The five destroyers were able to close without being seen and to launch their torpedoes. The darkness was suddenly lit by orange flame as the torpedoes struck home. When the destroyers retired without loss they left behind a scene of destruction. The *Fuso* was left listing and burning furiously, one destroyer was sunk and two more crippled and the *Yamashiro* had to struggle on with two of her magazines flooded.

Not long afterwards the second destroyer division attacked, sinking one of the disabled Japanese destroyers and putting yet another torpedo into the *Yamashiro*.

Despite these attacks Admiral Nishimura steamed on, being too determined to break into Leyte Gulf to worry about his sunk and damaged ships. Nishimura was only concerned about bringing his still powerful force amongst the American invasion fleet which would be helpless before the powerful guns of his remaining battleship and the heavy cruiser. He was completely unaware of Oldendorf's "reception committee"—this was made up of a line of six battleships, three heavy and two light cruisers on one flank and another heavy and two more light cruisers on the other, together with divisions of destroyers strung around the entrance to the trap.

The Japanese battleship *Yamato* pictured in 1941.

The Japanese ships came on, the *Yamashiro* leading, a destroyer *Shigure* on her starboard quarter and the *Mogami* following astern. They steamed straight into the 14- and 16-inch guns of the battleships and heavy cruisers. Oldendorf had accomplished the one great aim of every naval commander, he was crossing his enemy's 'T', as Togo had done years before at Tsushima.

Just before 4 a.m. the cruisers opened fire until, suddenly realizing that he was facing a mass of warships where he had only expected to find lightly armed assault and landing craft, Nishimura radioed to the *Fuso* to make full speed and join him. But the *Fuso* was well astern, and blazing.

Oldendorf's American battle-wagons now joined in, three of them equipped with the latest Mark-8 fire control radar; this meant that they could pinpoint their target in the darkness. These three, the *West Virginia, Tennessee* and *California,* fired six-gun salvoes with their 14-inch guns. The *Maryland* and *Mississippi* managed to join in by ranging on the other ships' splashes, whilst the *Pennsylvania,* with a poorer type of radar, did not fire at all.

The Japanese destroyer *Shigure,* a small target in the darkness, was hit only once, but her larger consort was continually struck by shells ranging from the 6-inch shells of the light cruisers to the armour-piercing shells of the battle-wagons. Both ships, nevertheless, fired back until the *Yamashiro*, ablaze from stem to stern, was struck by a pattern of torpedoes fired by two destroyers. Just before 4.15 a.m. she took on a frightful list then finally sank, taking Admiral Nishimura and most of his crew with her. A final salvo from the *Mississippi* sounded her death knell. It also marked the passing of the big gun type of naval warfare that had existed since the days of Francis Drake.

The Second Striking Force under Shima also sailed into the entrance to Surigao Strait and almost at once one of his light cruisers was torpedoed by a PT. Passing the fiercely burning *Fuso,* Shima realized that something had gone terribly wrong with the other Striking Force and he ordered his squadron to turn and retire. As his own ship the *Nachi* was turning it was run down by the crippled *Mogami* and badly damaged.

Admiral Oldendorf did not pursue the fleeing enemy. His main task was to protect the American landing forces. He also knew that another powerful Japanese fleet was in the vicinity. He ordered two of his light cruisers, the *Denver* and *Columbia* to follow for a while and "polish off enemy cripples". They did this by sinking a badly damaged destroyer, and then returned when they learned that the *Mogami* had been sunk by carrier-borne Avenger aircraft. Only the *Shigure* was able to escape to her base on Brunei Bay. She was the sole survivor of the world's last naval battle fought entirely between warships.

The final two battles of the Leyte Gulf were mainly between American "flat tops" (aircraft carriers) and Japanese warships which were aided by fanatical *kamikaze* (suicide) pilots which, like well-aimed shells, came roaring out of the sun in the hope of crashing on the decks of the U.S. warships before being shot down themselves.

The great victory of Leyte Gulf secured General MacArthur's communications and supply line. It was the prelude to the final defeat of the Japanese nation which surrendered unconditionally on August 14th, 1945, after atomic bombs had been dropped on Hiroshima and Nagasaki. With their detonation the age of the big gun was over.

For better or worse the world had moved into the nuclear age.

From top to bottom: A Japanese bomber nose-dives after being hit by American ack-ack fire; a Japanese torpedo plane bursts into flames after being hit by an American naval gun; and (*bottom*) batteries of American howitzers trained on the Japanese positions at Leyte.

Hiroshima after the dropping of the atomic bomb.

Nuclear destruction at Nagasaki.

Below The types of atomic bomb that wrought such devastation, the so-called "Little Boy" (*left*) of Hiroshima and "Fat Man" (*right*) of Nagasaki.

Chapter Nine

Current Trends

Past, Present and Future
The Atom Bomb

The effects of the first atom bomb to be dropped upon Hiroshima on August 6th, 1945, stunned the whole world. How could just one bomb, dropped from a single high-flying aircraft, cause such damage?

A Japanese journalist described what happened to his fellow countrymen immediately after the bomb had exploded: "Many were killed instantly, others lay writhing on the ground, screaming in agony from the intolerable pain of their burns. Everything standing upright in the way of the blast—walls, houses, factories, and other buildings—was annihilated and the debris spun round in a whirlwind and was carried up into the air . . . horses, dogs and cattle suffered the same fate as human beings. Every living thing was petrified in an attitude of indescribable suffering. . ."

The bomb that had been dropped was of 20 kilotons, that is to say, it had the explosive power of 20,000 tons of T.N.T. Nearly five square miles of the city were completely destroyed. Something like 80,000 people were killed and a further 50,000 injured. The dropping of a further atomic bomb on Nagasaki three days later was sufficient to end the war.

Despite the steady increase in larger and more powerful nuclear weapons, every war fought since 1945—in Korea, Indo-China, the Middle East and Vietnam—has been fought with conventional, non-nuclear weapons. It would seem that every nation with nuclear weapons, although well prepared for a full-scale holocaust, has played safe and fought limited wars with the same weapons that were used in World War Two, and before the first atom bomb was dropped.

Artillery

Today's guns, howitzers and mortars are of the type that have been in existence for many years and are still in every nation's armoury although there are weapons on the secret list which are capable of firing nuclear projectiles. One of these was the American 280-mm Gun M65, nicknamed "Atomic Annie". This was a very large gun, redesigned to provide a long-range weapon capable of firing a nuclear shell. Its size made it impracticable and it has never been used. The largest calibre gun of World War Two was the American "Little David", an RML with a calibre of 36 inches. It, too, has never been used in action.

A comparatively new development in heavy field guns is the SP or self-propelled gun, a weapon which developed naturally from the tank. The first of these was built in 1917 and consisted of the chassis of a Mark 1 tank and a 6-inch howitzer, with its wheels removed, placed upon it. For a time the idea was not pursued but when World War Two began, the possibilities of the SP were rediscovered and many of them were built. Indeed, by the end of the war the German army had more self-propelled guns (*Panzerjäger*) than tanks.

Britain also has had a number of successful SP guns including the *Sexton,* armed with a 25-pounder gun and the *Archer,* with a 17-pounder mounted on a Valentine tank chassis. The *Abbot,* the SP operational with the British army today, is regarded as the best SP gun now in use.

The Americans too have had all kinds of excellent SP guns. The 105-mm on a half-track and the 155-mm Howitzer M41 are good examples. Their largest was the *King Kong,* a 240-mm howitzer.

Russia has also produced some powerful SP guns. Two, mounted on the Josef Stalin tank chassis, the JSU 122 (122-mm) and the JSU 152 (152-mm).

Today, most of the world's larger field artillery is self-propelled, with the exception of the U.S.S.R. which still relies upon the two-wheeled towing chassis.

Naval Guns

By the end of World War Two battleships had become obsolete and aircraft carriers taken their place.

An Israeli 203-mm. self-propelled howitzer, made in the United States. It has a range of 16·8 kilometres.

These, in turn, are slowly being phased out in favour of submarines, which, nuclear-powered and armed with powerful ballistic missiles, can fire their S.L.B.M. (or submarine-launched ballistic missiles) without surfacing.

It is unusual for warships today to carry guns larger than the 8-inch. Indeed the world's largest carrier, the American 85,350-ton *Enterprise,* has no guns and no missiles. She relies upon her 100 or so aircraft and her escorts for protection. The *Forrestal, America* and *Midway*, three other huge American carriers carry only four 5-inch guns each, mainly for use against aircraft.

The world's first atomic-powered submarine, the American *Nautilus*, launched in 1954.

Rockets . . .

The principle of the rocket for war purposes is by no means new. Rocket Cavalry and Infantry and Rocket Cars were part of the British army as early as 1811, the invention of Colonel William Congreve. Of all the nations who had an opportunity to see their potential, only Germany went ahead with experiments. They built a 12-inch diameter and 4½-ft long rocket, the *A1* in 1933, improving on this until the outbreak of World War Two when they had developed the *A3*, a rocket 25 ft long and 2½ ft in diameter.

Its successor was the *A4* or, more familiarly, the *V2* rocket. Prior to this, Germany used another unmanned missile, the *V1,* known to British people in wartime as the "Doodle-bug". The first of these landed in Brixton, London, on 13 June, 1944, a week after D-Day. Within the next five weeks more than 3,000 fell in and around London, causing great damage and loss of life. The *V1* carried about a ton of explosive and had a small propellor driven by the passage of the bomb through the air. When the *V1* had flown a pre-set distance from its launching site to arrive over London, the engine cut out and the bomb glided silently down to earth to explode violently. These unmanned missiles were finally overcome by the use of fighter aircraft, AA guns and balloons.

The next threat came from the German long-range rocket, or *V2*. This was a far larger weapon than the *V1,* being twelve tons in weight with a one-ton warhead, and a range of more than 200 miles. It was jet-driven, the thrust being the result of the combustion of alcohol and liquid oxygen, forced into a chamber by a 1,000 horsepower pump. When launched the rocket rose vertically for 6 miles then its automatic controls took over to carry it up to a total height of 50 miles. The missile then raced towards the earth at a speed in excess of 4,000 miles an hour. This was faster than the speed of sound and consequently there was no chance to escape—if you heard the bang you could congratulate yourself . . . you were still alive!

There was no defence against these. The terrible threat was not removed until the Royal Air Force had flattened many of the launching sites with bombs and the advancing Allied armies had overrun the rest.

. . . and Missiles

Missiles fall into two categories, ballistic and non-ballistic. Broadly speaking, they may be defined as *ballistic*—a missile fired from a gun or launcher using a propellant (the common shell is a ballistic missile, "ballistic" meaning projectile) and *non-ballistic*—a missile which contains its own power source and which, in many cases, can be guided to its target.

Ballistic missiles have developed from the marriage of the *V2* and the nuclear warhead. They are launched from massive, concrete structures deep in the ground known as silos; from submerged submarines; from surface ships, and from special mobile carriers. These include the Intercontinental Ballistic Missiles (I.C.B.M.) with a range of more than 8,000 miles, more than the diameter of the earth. The first I.C.B.M. was the American *Atlas* which made its first successful flight in 1957. The latest is the U.S. *Minuteman,* a three-stage missile some 54 feet long and 60,000 lbs in weight. It can travel more than 6,000 miles at a speed of Mach 22, which is 22 times the speed of sound (or nearly 17,000 miles an hour). It has a two-megaton warhead which means that each *Minuteman* has the detonating power equal to *two million* tons of the high explosive T.N.T.

The largest known, however, is the *Titan II* which has a 9,000-mile range and a warhead of more than five megatons, although it is believed that both America and Russia have I.C.B.M.s with warheads of more than 25 megatons (or 25 million tons). By comparison, the entire weight of bombs dropped by the Allies, in every theatre of operation in World War Two was less than four million tons!

Most missiles, however, fall into the non-ballistic category and are classified as:

AIR-TO-AIR (A.A.M.) carried in aircraft to fire at other aircraft (e.g. the British *Firestreak* and *Red Top*).

AIR-TO-SURFACE (A.S.M.) or missiles launched from aircraft at land or seaborne targets (e.g. the British *Blue Steel*).

SURFACE-TO-AIR (S.A.M.) which are mainly anti-aircraft missiles (e.g. British *Bloodhound* and *Thunderbird,* and for British naval vessels, the *Seacat, Sea Dart* and *Seaslug*). Some in this range are designed to intercept enemy missiles outside the atmosphere and may be regarded as anti-missile missiles (e.g. the U.S. *Sprint* and Soviet *Galosh.*)

SURFACE-TO-SURFACE missiles can be either ballistic or non-ballistic. Some are huge, for long-range bombardment; some are small, non-nuclear missiles for use against tanks (e.g. the British *Swingfire*).

SUBMARINE AND ANTI-SUBMARINE MISSILES. Allied ballistic-missile submarines are known as *Polaris* submarines from the type of missile they fire. The *Polaris* missile has a range of up to 2,875 miles and carries a 0·7 megaton warhead. British nuclear-powered ballistic missile carrying submarines are of the *Resolution* class.

Anti-submarine missiles are fitted with sonar (sound) detectors and a computer enabling it to track and then aim itself at the target. (e.g. the British *Ikara,* the U.S. *Asroc* and *Subroc*).

The destructive power of a nuclear warhead has been referred to as being measured in megatons (or million tons). In modern parlance, however, megatons are no longer used. The expression now used is OVERKILL, which represents the capacity of a nation armed with nuclear equipment to

Above The giant American Titan on its launching pad.
Top right A British Swingfire launching a long-range anti-tank missile.
Centre Lift-off of a British Thunderbird missile during practice.
Bottom The horror of the atomic mushroom cloud. The shape of things to come . . . ?

kill everyone alive when all their missiles are detonated. The capacity of the United States of America, for example, is considered to be at least Overkill 7, which means that her current available nuclear weapons could kill seven times the number of people living on this planet, at this moment of time.

And what a frightening thought that is.

Further Reading

Books

Great Military Sieges, Vegio Melegari (New English Libary)

The Art of Warfare on Land, D. Chandler (Hamlyn)

The Guinness History of Land Warfare, Kenneth Macksey (Guinness Superlatives)

Dictionary of Modern War, Edward Luttwak (Allen Lane: The Penguin Press)

A History of Weaponry, Courtlandt Canby (Prentice-Hall International)

NATO Armour, Bryan Perrett (Ian Allen)

Great Military Battles, Cyril Falls (edit) (Weidenfeld & Nicolson)

Guns, Dudley Pope (Weidenfeld & Nicolson)

The Machinery of War, Peter Young (Hart-Davis, MacGibbon)

The Two-Ocean War, Samuel E. Morison (Atlantic Monthly Press)

Great Sea Battles, Oliver Warner (Weidenfeld & Nicolson)

Artillery through the Ages, H. C. B. Rogers (Seeley Service)

Artillery, John Batchelor and Ian Hogg (Macdonald)

Part Works

The First World War (Purnells)

The Second World War (Purnells)

The Sea (Orbis)

History of Ships (New English Library)

The War Illustrated (Amalgamated Press)

Index

Figures in bold refer to illustrations

PICTURE CREDITS

The author and publishers wish to thank the following for permission to reproduce copy-right illustrations appearing on the pages mentioned: Imperial War Museum, *half title page, frontispiece*, 23, 26 *centre and bottom*, 27 *centre and bottom*, 29 *centre and bottom*, 30, 33, 36 *bottom*, 37 *bottom*, 38, 39 *centre*, 40, 48, 50, 51 *top and bottom*, 53, 55, 57, 60, 61, 62 *bottom left and right*, 64, 65, 67, 70 *top*, 72–3, 74, 75, 76–7, 79, 81, 82, 87 *bottom*; Radio Times Hulton Picture Library, 4, 6 *bottom right*, 8, 9, 12–13, 15 *bottom right*, 17, 21, 22 *bottom*, 39 *bottom*, 44–5, 46 *top and bottom*, 62 *top*; The Mansell Collection, *endpapers*, 10, 11, 14 *top left and right*; Novosti Press Agency, *title page*, 66; British Museum, 6 *top left and bottom left*, 7; Department of the Environment Edinburgh, 19; Keystone Press Agency, 70 *bottom*, 84, 85, 87 *top left*; The Science Museum, 24; British Aircraft Corporation, 87 *top right and centre*.